Pleasures of the Telescope

by Garrett Servis

PREFACE

By the introduction of a complete series of star maps, drawn specially for the use of the amateur and distributed through the body of the work, thus facilitating consultation, it is believed that this book makes a step in advance of its predecessors. The maps show all of the stars visible to the naked eye in the regions of sky represented, and, in addition, some stars that can only be seen with optical aid. The latter have been placed in the maps as guide posts in the telescopic field to assist those who are searching for faint and inconspicuous objects referred to in the text. As the book was not written for those who possess the equipment of an observatory, with telescopes driven by clockwork and provided with graduated circles, right ascensions and declinations are not given. All of the telescopic phenomena described are, however, represented in the maps. Star clusters are indicated by a conventional symbol, and nebulaeby a little white circle; while a small cross serves to mark the places where notable new stars have appeared. The relative magnitudes of the stars are approximately shown by the dimensions of their symbols in the maps, the smaller stars being represented by white dots and the larger by star-shaped figures.

In regard to binary stars, it should be remembered that, in many cases, their distances and angles of position change so rapidly that any statement concerning them remains valid only for a few years at the most. There is also much confusion among the measurements announced by different authorities. In general, the most recent measurements obtainable in 1900 are given in the text, but the observer who wishes to study close and rapid binaries will do well to revise his information about them as frequently as possible. An excellent list of double stars kept up to date, will be found in the annual Companion to the Observatory, published in London.

In the lunar charts the plan of inserting the names of the principal formations has been preferred to that usually followed, of indicating them only by numbers, accompanied by a key list. Even in the most detailed charts of the moon only a part of what is visible with telescopes can be shown, and the representation, at best, must be merely approximate. It is simply a question of what to include and what to omit; and in the present case the probable needs of the amateur observer have governed the selection-- readiness and convenience of reference being the chief aim.

It should, perhaps, be said here that the various chapters composing this book--like those of "Astronomy with an Opera-glass"--were, in their original form, with the single exception of Chapter IX, published in Appletons' Popular Science Monthly. The author, it is needless to say, was much gratified by the expressed wish of many readers that these scattered papers should be revised and collected in a more permanent form. As bearing upon the general subject of the book, a chapter has been added, at the end, treating on the question of the existence of planets among the stars. This also first appeared in the periodical above mentioned.

In conclusion, the author wishes for his readers as great a pleasure in the use of the telescope as he himself has enjoyed.

G. P. S.

BOROUGH OF BROOKLYN, NEW YORK, January, 1901.

CONTENTS

Significance of Dr. See's observations--Why our telescopes do not show planets circling around distant suns--Reasons for thinking that such planets may exist--The bearing of stellar evolution on the question.

PLEASURES OF THE TELESCOPE

CHAPTER I

THE SELECTION AND TESTING OF A GLASS

"O telescope, instrument of much knowledge, more precious than any scepter! Is not he who holds thee in his hand made king and lord of the works of God?"--JOHN KEPLER.

If the pure and elevated pleasure to be derived from the possession and use of a good telescope of three, four, five, or six inches aperture were generally known, I am certain that no instrument of science would be more commonly found in the homes of intelligent people. The writer, when a boy, discovered unexpected powers in a pocket telescope not more than fourteen inches long when extended, and magnifying ten or twelve times. It became his dream, which was afterward realized, to possess a more powerful telescope, a real astronomical glass, with which he could see the beauties of the double stars, the craters of the moon, the spots on the sun, the belts and satellites of Jupiter, the rings of Saturn, the extraordinary shapes of the nebulae the crowds of stars in the Milky Way, and the great stellar clusters. And now he would do what he can to persuade others, who perhaps are not aware how near at hand it lies, to look for themselves into the wonder-world of the astronomers.

There is only one way in which you can be sure of getting a good telescope. First, decide how large a glass you are to have, then go to a maker of established reputation, fix upon the price you are willing to pay-- remembering that good work is never cheap--and finally see that the instrument furnished to you answers the proper tests for a telescope of its size. There are telescopes and telescopes. Occasionally a rare combination of perfect homogeneity in the material, complete harmony between the two kinds of glass of which the objective is composed, and lens surfaces whose curves are absolutely right, produces a telescope whose owner would part with his last dollar sooner than with it. Such treasures of the lens-maker's art can not, perhaps, be commanded at will, yet, they are turned out with increasing frequency, and the best artists are generally able, at all times, to approximate so closely to perfection that any shortcoming may be disregarded.

In what is said above I refer, of course, to the refracting telescope, which is the form of instrument that I should recommend to all amateurs in preference to the reflector. But, before proceeding further, it may be well to recall briefly the principal points of difference between these two kinds of telescopes. The purpose of a telescope of either description is, first, to form an image of the object looked at by concentrating at a focus the rays of light proceeding from that object. The refractor achieves this by means of a carefully shaped lens, called the object glass, or objective. The reflector, on the other hand, forms the image at the focus of a concave mirror.

A very pretty little experiment, which illustrates these two methods of forming an optical image, and, by way of corollary, exemplifies the essential difference between refracting and reflecting telescopes, may be performed by any one who possesses a reading glass and a magnifying hand mirror. In a room that is not too brightly illuminated pin a sheet of white paper on the wall opposite to a window that, by preference, should face the north, or away from the position of the sun. Taking first the reading glass, hold it between the window and the wall parallel to the sheet of paper, and a foot or more distant from the latter. By moving it to and fro a little you will be able to find a distance, corresponding to the focal length of the lens, at which a picture of the window is formed on the paper. This picture, or image, will be upside down, because the rays of light cross at the focus. By moving the glass a little closer to the wall you will cause the picture of the window to become indistinct, while a beautiful image of the houses, trees, or other objects of the outdoor world beyond, will be formed upon the paper. We thus learn that the distance of the image from the lens varies with the distance of the object whose image is formed. In precisely a similar manner an image is formed at the focus of the object glass of a refracting telescope.

Take next your magnifying or concave mirror, and detaching the sheet of paper from the wall, hold it nearly in front of the mirror between the latter and the window. When you have adjusted the distance to the focal length of the mirror, you will see an image of the window projected upon the paper, and by varying the distance, as before, you will be able to produce, at will, pictures of nearer or more remote objects. It is in this way that images are formed at the focus of the mirror of a reflecting telescope.

Now, you will have observed that the chief apparent difference between these two methods of forming an image of distant objects is that in the first case the rays of light, passing through the transparent lens, are brought to a focus on the side opposite to that where the real object is, while in the second case the rays, being reflected from the brilliant surface of the opaque mirror, come to a focus on the same side as that on which the object itself is. From this follows the most striking difference in the method of using refracting and reflecting telescopes. In the refractor the observer looks toward the object; in the reflector he looks away from it. Sir William Herschel made his great discoveries with his back to the sky. He used reflecting telescopes. This principle, again, can be readily illustrated by means of our simple experiment with a reading glass and a magnifying mirror. Hold the reading glass between the eye and a distant object with one hand, and with the other hand place a smaller lens such as a pocket magnifier, near the eye, and in line with the reading glass. Move the two carefully until they are at a distance apart equal to the sum of the focal lengths of the lenses, and you will see a magnified image of the distant object. In other words, you have constructed a simple refracting telescope. Then take the magnifying mirror, and, turning your back to the object to be looked at, use the small lens as before--that is to say, hold it between your eye and the mirror, so that its distance from the latter is equal to the sum of the focal lengths of the mirror and the lens, and you will see again a magnified image of the distant object. This time it is a reflecting telescope that you hold in your hands.

The magnification of the image reminds us of the second purpose which is subserved by a telescope. A telescope, whether refracting or reflecting, consists of two essential parts, the first being a lens, or a mirror, to form an image, and the second a microscope, called an eyepiece, to magnify the image. The same eyepieces will serve for either the reflector or the refractor. But in order that the magnification may be effective, and serve to reveal what could not be seen without it, the image itself must be as nearly perfect as possible; this requires that every ray of light that forms the image shall be brought to a point in the image precisely corresponding to that from which it emanates in the real object. In reflectors this is effected by giving a parabolic form to the concave surface of the mirror. In refractors there is a twofold difficulty to be overcome. In the first place, a lens with spherical surfaces does not bend all the rays that pass through it to a focus at precisely the same distance. The rays that pass near the outer edge of the lens have a shorter focus than that of the rays which pass near the center of the lens; this is called spherical aberration. A similar phenomenon occurs with a concave mirror whose surface is spherical. In that case, as we have seen, the difficulty is overcome by giving the mirror a parabolic instead of a spherical form. In an analogous way the spherical aberration of a lens can be corrected by altering its curves, but the second difficulty that arises with a lens is not so easily disposed of: this is what is called chromatic aberration. It is due to the fact that the rays belonging to different parts of the spectrum have different degrees of refrangibility, or, in other words, that they come to a focus at different distances from the lens; and this is independent of the form of the lens. The blue rays come to a focus first, then the yellow, and finally the red. It results from this scattering of the spectral rays along the axis of the lens that there is no single and exact focus where all meet, and that the image of a star, for instance, formed by an ordinary lens, even if the spherical aberration has been corrected, appears blurred and discolored. There is no such difficulty with a mirror, because there is in that case no refraction of the light, and consequently no splitting up of the elements of the spectrum.

In order to get around the obstacle formed by chromatic aberration it is necessary to make the object glass of a refractor consist of two lenses, each composed of a different kind of glass. One of the most interesting facts in the history of the telescope is that Sir Isaac Newton could see no hope that chromatic aberration would be overcome, and accordingly turned his attention to the improvement of the reflecting telescope and devised a form of that instrument which still goes under his name. And even after Chester More Hall in 1729, and John Dollond in 1757, had shown that chromatic aberration could be nearly eliminated by the combination of a flint-glass lens with one of crown glass, William Herschel, who began his observations in 1774, devoted his skill entirely to the making of reflectors, seeing no prospect of much advance in the power of refractors.

A refracting telescope which has been freed from the effects of chromatic aberration is called achromatic. The principle upon which its construction depends is that by combining lenses of different dispersive power the separation of the spectral colors in the image can be corrected while the convergence of the rays of light toward a focus is not destroyed. Flint glass effects a greater dispersion than crown glass nearly in the ratio of three to two. The chromatic combination consists of a convex lens of crown backed by a concave, or plano-concave, lens of flint. When these two lenses are made of focal lengths which are directly proportional to their dispersions, they give a practically colorless image at their common focus. The skill of the telescope-maker and the excellence of his work depend upon the selection of the glasses to be combined and his manipulation of the curves of the lenses.

Now, the reader may ask, "Since reflectors require no correction for color dispersion, while that correction is only approximately effected by the combination of two kinds of lenses and two kinds of glass in a refractor, why is not the reflector preferable to the refractor?"

The answer is, that the refractor gives more light and better definition. It is superior in the first respect because a lens transmits more light than a mirror reflects. Professor Young has remarked that about eighty-two per cent of the light reaches the eye in a good refractor, while "in a Newtonian reflector, in average condition, the percentage seldom exceeds fifty per cent, and more frequently is lower than higher." The superiority of the refractor in regard to definition arises from the fact that any distortion at the surface of a mirror affects the direction of a ray of light three times as much as the same distortion would do at the surface of a lens. And this applies equally both to permanent errors of curvature and to temporary distortions produced by strains and by inequality of temperature. The perfect achromatism of a reflector is, of course, a great advantage, but the chromatic aberration of refractors is now so well corrected that their inferiority in that respect may be disregarded. It must be admitted that reflectors are cheaper and easier to make, but, on the other hand, they require more care, and their mirrors frequently need resilvering, while an object glass with reasonable care never gets seriously out of order, and will last for many a lifetime.

Enough has now, perhaps, been said about the respective properties of object glasses and mirrors, but a word should be added concerning eyepieces. Without a good eyepiece the best telescope will not perform well. The simplest of all eyepieces is a single double-convex lens. With such a lens the magnifying power of the telescope is measured by the ratio of the focal length of the objective to that of the eye lens. Suppose the first is sixty inches and the latter half an inch; then the magnifying power will be a hundred and twenty diameters--i. e., the disk of a planet, for instance, will be enlarged a hundred and twenty times along each diameter, and its area will be enlarged the square of a hundred and twenty, or fourteen thousand four hundred times. But in reckoning magnifying power, diameter, not area, is always considered. For practical use an eyepiece composed of an ordinary single lens is seldom advantageous, because good definition can only be obtained in the center of the field. Lenses made according to special formul? however, and called solid eyepieces, give excellent results, and for high powers are often to be preferred to any other. The eyepieces usually furnished with telescopes are, in their essential principles, compound microscopes, and they are of two descriptions, "positive" and "negative." The former generally goes under the name of its inventor, Ramsden, and the latter is named after great Dutch astronomer, Huygens. The Huygens eyepiece consists of two plano-convex lenses whose focal lengths are in the ratio of three to one. The smaller lens is placed next to the eye. Both lenses have their convex surfaces toward the object glass, and their distance apart is equal to half the sum of their focal lengths. In this kind of eyepiece the image is formed between the two lenses, and if the work is properly done such an eyepiece is achromatic. It is therefore generally preferred for mere seeing purposes. In the Ramsden eyepiece two plano-convex lenses are also used, but they are of equal focal length, are placed at a distance apart equal to two thirds of the focal length of either, and have their convex sides facing one another. With such an eyepiece the image viewed is beyond the farther or field lens instead of between the two lenses, and as this fact renders it easier to adjust wires or lines for measuring purposes in the focus of the eyepiece, the Ramsden construction is used when a micrometer is to be employed. In order to ascertain the magnifying power which an eyepiece gives when applied to a telescope it is necessary to know the equivalent, or combined, focal length of the two lenses. Two simple rules, easily remembered, supply the means of ascertaining this. The equivalent focal length of a negative or Huygens eyepiece is equal to half the focal length of the larger or field lens. The equivalent focal length of a positive or Ramsden eyepiece is equal to three fourths of the focal length of either of the lenses. Having ascertained the equivalent focal length of the eyepiece, it is only necessary to divide it into the focal length of the object glass (or mirror) in order to know the magnifying power of your telescope when that particular eyepiece is in use.

A first-class object glass (or mirror) will bear a magnifying power of one hundred to the inch of aperture when the air is in good condition--that is, if you are looking at stars. If you are viewing the moon, or a planet, better results will always be obtained with lower powers--say fifty to the inch at the most. And under ordinary atmospheric conditions a power of from fifty to seventy-five to the inch is far better for stars than a higher power. With a five-inch telescope that would mean from two hundred and fifty to three hundred and seventy-five diameters, and such powers should only be applied for the sake of separating very close double stars. As a general rule, the lowest power that will distinctly show what you desire to see gives the best results. The experienced observer never uses as high powers as the beginner does. The number of eyepieces purchased with a telescope should never be less than three--a very low power--say ten to the inch; a very high power, seventy-five or one hundred to the inch, for occasional use; and a medium power--say forty to the inch--for general use. If you can afford it, get a full battery of eyepieces--six or eight, or a dozen--for experience shows that different objects require different powers in order to be best seen, and, moreover, a slight change of power is frequently a great relief to the eye.

There is one other thing of great importance to be considered in purchasing a telescope--the mounting. If your glass is not well mounted on a steady and easily managed stand, you might better have spent your money for something more useful. I have endured hours of torment while trying to see stars through a telescope that was shivering in the wind and dancing to every motion of the bystanders, to say nothing of the wriggling contortions caused by the application of my own fingers to the focusing screw. The best of all stands is a solid iron pillar firmly fastened into a brick or stone pier, sunk at least four feet in the ground, and surmounted by a well-made equatorial bearing whose polar axis has been carefully placed in the meridian. It can be readily protected from the weather by means of a wooden hood or a rubber sheet, while the tube of the telescope may be kept indoors, being carried out and placed on its bearing only when observations are to be made. With such a mounting you can laugh at the observatories with their cumbersome domes, for the best of all observatories is the open air. But if you dislike the labor of carrying and adjusting the tube every time it is used, and are both fond of and able to procure luxuries, then, after all, perhaps, you had better have the observatory, dome, draughts and all.

The next best thing in the way of a mounting is a portable tripod stand. This may be furnished either with an equatorial bearing for the telescope, or an altazimuth arrangement which permits both up-and-down and horizontal motions. The latter is cheaper than the equatorial and proportionately inferior in usefulness and convenience. The essential principle of the equatorial bearing is motion about two axes placed at right angles to one another. When the polar axis is in the meridian, and inclined at an angle equal to the latitude of the place, the telescope can be moved about the two axes in such a way as to point to any quarter of the sky, and the motion of a star, arising from the earthy rotation, can be followed hour after hour without disturbing the instrument. When thus mounted, the telescope may be driven by clockwork, or by hand with the aid of a screw geared to a handle carrying a universal joint.

And now for testing the telescope. It has already been remarked that the excellence of a telescope depends upon the perfection of the image formed at the focus of the objective. In what follows I have only a refractor in mind, although the same principles would apply to a reflector. With a little practice anybody who has a correct eye can form a fair judgment of the excellence of a telescopic image. Suppose we have our telescope steadily mounted out of doors (if you value your peace of mind you will not try to use a telescope pointed out of a window, especially in winter), and suppose we begin our observations with the pole star, employing a magnifying power of sixty or seventy to the inch. Our first object is to see if the optician has given us a good glass. If the air is not reasonably steady we had better postpone our experiment to another night, because we shall find that the star as seen in the telescope flickers and "boils," and behaves in so extraordinary a fashion that there is no more definition in the image than there is steadiness in a bluebottle buzzing on a window pane. But if the night is a fine one the star image will be quiescent, and then we may note the following particulars: The real image is a minute bright disk, about one second of arc in diameter if we are using a four-and-a-half or five-inch telescope, and surrounded by one very thin ring of light, and the fragments, so to speak, of one or possibly two similar rings a little farther from the disk, and visible, perhaps, only by glimpses. These "diffraction rings" arise from the undulatory nature of light, and their distance apart as well as the diameter of the central disk depend upon the length of the waves of light. If the telescope is a really good one, and both object glass and eyepiece are properly adjusted, the disk will be perfectly round, slightly softer at the edge, but otherwise equally bright throughout; and the ring or rings surrounding it will be exactly concentric, and not brighter on one side than on another. Even if our telescope were only two inches or two inches and a half in aperture we should at once notice a little bluish star, the mere ghost of a star in a small telescope, hovering near the polar star. It is the celebrated "companion," but we shall see it again when we have more time to study it. Now let us put the star out of focus by turning the focusing screw. Suppose we turn it in such a way that the eyepiece moves slightly outside the focus, or away from the object glass. Very beautiful phenomena immediately begin to make their appearance. A slight motion outward causes the little disk to expand perceptibly, and just as this expansion commences, a bright-red point appears at the precise center of the disk. But, the outward motion continuing, this red center disappears, and is replaced by a blue center, which gradually expands into a sort of flare over the middle of the disk. The disk itself has in the mean time enlarged into a series of concentric bright rings, graduated in luminosity with beautiful precision from center toward circumference. The outermost ring is considerably brighter, however, than it would be if the same gradation applied to it as applies to the inner rings, and it is surrounded, moreover, on its outer edge by a slight flare which tends to increase its apparent width. Next let us return to the focus and then move the eyepiece gradually inside the focal point or plane. Once more the star disk expands into a series of circles, and, if we except the color phenomena noticed outside the focus, these circles are precisely like those seen before in arrangement, in size, and in brightness. If they were not the same, we should pronounce the telescope to be imperfect. There is one other difference, however, besides the absence of the blue central flare, and that is a faint reddish edging around the outer ring when the expansion inside the focus is not carried very far. Upon continuing to move the eyepiece inside or outside the focus we observe that the system of rings becomes larger, while the rings themselves rapidly increase in number, becoming at the same time individually thinner and fainter.

By studying the appearance of the star disk when in focus and of the rings when out of focus on either side, an experienced eye can readily detect any fault that a telescope may have. The amateur, of course, can only learn to do this by considerable practice. Any glaring and serious fault, however, will easily make itself manifest. Suppose, for example, we observe that the image of a star instead of being perfectly round is oblong, and that a similar defect appears in the form of the rings when the eyepiece is put out of focus. We know at once that something is wrong; but the trouble may lie either in the object glass, in the eyepiece, in the eye of the observer himself, or in the adjustment of the lenses in the tube. A careful examination of the image and the out-of-focus circles will enable us to determine with which of these sources of error we have to deal. If the star image when in focus has a sort of wing on one side, and if the rings out of focus expand eccentrically, appearing wider and larger on one side than on the other, being at the same time brightest on the least expanded side, then the object glass is probably not at right angles to the axis of the tube and requires readjustment. That part of the object glass on the side where the rings appear most expanded and faintest needs to be pushed slightly inward. This can be effected by means of counterscrews placed for that purpose in or around the cell. But it, after we have got the object glass properly squared to the axis of the tube or the line of sight, the image and the ring system in and out of focus still appear oblong, the fault of astigmatism must exist either in the objective, the eyepiece, or the eye. The chances are very great that it is the eye itself that is at fault. We may be certain of this if we find, on turning the head so as to look into the telescope with the eye in different positions, that the oblong image turns with the head of the observer, keeping its major axis continually in the same relative position with respect to the eye. The remedy then is to consult an oculist and get a pair of cylindrical eyeglasses. If the oblong image does not turn round with the eye, but does turn when the eyepiece is twisted round, then the astigmatism is in the latter. If, finally, it does not follow either the eye or the eyepiece, it is the objective that is at fault.

But instead of being oblong, the image and the rings may be misshapen in some other way. If they are three-cornered, it is probable that the object glass is subjected to undue pressure in its cell. This, if the telescope has been brought out on a cool night from a warm room, may arise from the unequal contraction of the metal work and the glass as they cool off. In fact, no good star image can be got while a telescope is assuming the temperature of the surrounding atmosphere. Even the air inclosed in the tube is capable of making much trouble until its temperature has sunk to the level of that outside. Half an hour at least is required for a telescope to adjust itself to out-of-door temperature, except in the summer time, and it is better to allow an hour or two for such adjustment in cold weather. Any irregularity in the shape of the rings which persists after the lenses have been accurately adjusted and the telescope has properly cooled may be ascribed to imperfections, such as veins or spots of unequal density in the glass forming the objective.

The spherical aberration of an object glass may be undercorrected or overcorrected. In the former case the central rings inside the focus will appear faint and the outer ones unduly strong, while outside the focus the central rings will be too bright and the outer ones too feeble. But if the aberration is overcorrected the central rings will be overbright inside the focus and abnormally faint outside the focus.

Assuming that we have a telescope in which no obvious fault is discernible, the next thing is to test its powers in actual work. In what is to follow I shall endeavor to describe some of the principal objects in the heavens from which the amateur observer may expect to derive pleasure and instruction, and which may at the same time serve as tests of the excellence of his telescope. No one should be deterred or discouraged in the study of celestial objects by the apparent insignificance of his means of observation. The accompanying pictures of the planet Mars may serve as an indication of the fact that a small telescope is frequently capable of doing work that appears by no means contemptible when placed side by side with that of the greater instruments of the observatories.

CHAPTER II

IN THE STARRY HEAVENS

"Now constellations, Muse, and signs rehearse; In order let them sparkle in thy verse."--MANILIUS.

Let us imagine ourselves the happy possessors of three properly mounted telescopes of five, four, and three inches aperture, respectively. A fine midwinter evening has come along, the air is clear, cool, and steady, and the heavens, of that almost invisible violet which is reserved for the lovers of celestial scenery, are spangled with stars that hardly twinkle. We need not disturb our minds about a few thin clouds here and there floating lazily in the high air; they announce a change of weather, but they will not trouble us to-night.

Which way shall we look? Our eyes will answer the question for us. However we may direct them, they instinctively return to the south, and are lifted to behold Orion in his glory, now near the meridian and midway to the zenith, with Taurus shaking the glittering Pleiades before him, and Canis Major with the flaming Dog Star following at his heels.

Not only is Orion the most brilliant of all constellations to the casual star-gazer, but it contains the richest mines that the delver for telescopic treasures can anywhere discover. We could not have made a better beginning, for here within a space of a few square degrees we have a wonderful variety of double stars and multiple stars, so close and delicate as to test the powers of the best telescopes, besides a profusion of star-clusters and nebulae including one of the supreme marvels of space, the Great Nebula in the Sword.

Our star map No. 1 will serve as a guide to the objects which we are about to inspect. Let us begin operations with our smallest telescope, the three-inch. I may remark here that, just as the lowest magnifying power that will clearly reveal the object looked for gives ordinarily better results than a higher power, so the smallest telescope that is competent to show what one wishes to see is likely to yield more satisfaction, as far as that particular object is concerned, than a larger glass. The larger the object glass and the higher the power, the greater are the atmospheric difficulties. A small telescope will perform very well on a night when a large one is helpless.

Turn the glass upon beta (Rigel), the white first-magnitude star in Orion's left foot. Observe whether the image with a high power is clear, sharp, and free from irregular wisps of stray light. Look at the rings in and out of focus, and if you are satisfied with the performance, try for the companion. A good three-inch is certain to show it, except in a bad state of the atmosphere, and even then an expert can see it, at least by glimpses. The companion is of the ninth magnitude, some say the eighth, and the distance is about 9.5", angle of position (hereafter designated by p.) 199?[1] Its color is blue, in decided contrast with the white light of its great primary. Sir John Herschel, however, saw the companion red, as others have done. These differences are doubtless due to imperfections of the eye or the telescope. In 1871 Burnham believed he had discovered that the companion was an exceedingly close double star. No one except Burnham himself succeeded in dividing it, and he could only do so at times. Afterward, when he was at Mount Hamilton, he tried in vain to split it with the great thirty-six-inch telescope, in 1889, 1890, and 1891. His want of success induced him to suggest that the component stars were in rapid motion, and so, although he admitted that it might not be double after all, he advised that it should be watched for a few years longer. His confidence was justified, for in 1898 Aitken, with the Lick telescope, saw and measured the distance of the extremely minute companion--distance 0.17", p. 177?

[1] The angle of position measures the inclination to the meridian of a line drawn between the principal star and its companion; in other words, it shows in what direction from the primary we must look for the companion. It is reckoned from 0?up to 360? beginning at the north point and passing around by east through south and west to north again. Thus, if the angle of position is 0?or 360? the companion is on the north side of the primary; if the angle is 90? the companion is to the east; if 180? to the south; if 270? to the west, and so for intermediate angles. It must be remembered, however, that in the field of the telescope the top is south and the bottom north, unless a prism is used, when directions become complicated. East and west can be readily identified by noticing the motion of a star through the field; it moves toward the west and from the east.

Rigel has been suspected of a slight degree of variability. It is evidently a star of enormous actual magnitude, for its parallax escapes trustworthy measurement. It can only be ranked among the very first of the light-givers of the visible universe. Spectroscopically it belongs to a peculiar type which has very few representatives among the bright stars, and which has been thus described: "Spectra in which the hydrogen lines and the few metallic lines all appear to be of equal breadth and sharp definition." Rigel shows a line which some believe to represent magnesium; but while it has iron lines in its spectrum, it exhibits no evidence of the existence of any such cloud of volatilized iron as that which helps to envelop the sun.

For another test of what the three-inch will do turn to zeta, the lower, or left-hand, star in the Belt. This is a triple, the magnitudes being second, sixth, and tenth. The sixth-magnitude star is about 2.5" from the primary, p. 149? and has a very peculiar color, hard to describe. It requires careful focusing to get a satisfactory view of this star with a three-inch telescope. Use magnifying powers up to two hundred and fifty diameters. With our four-inch the star is much easier, and the five-inch shows it readily with a power of one hundred. The tenth-magnitude companion is distant 56", p. 8? and may be glimpsed with the three-inch. Upon the whole, we shall find that we get more pleasing views of zeta Orionis with the four-inch glass.

Just to the left of zeta, and in the same field of view with a very low power, is a remarkable nebula bearing the catalogue number 1227. We must use our five-inch on this with a low power, but with zeta out of the field in order to avoid its glare. The nebula is exceedingly faint, and we can be satisfied if we see it simply as a hazy spot, although with much larger telescopes it has appeared at least half a degree broad. Tempel saw several centers of condensation in it, and traced three or four broad nebulous streams, one of which decidedly suggested spiral motion.

The upper star in the Belt, delta, is double; distance, 53", p. 360? magnitudes, second and seventh very nearly; colors, white and green or blue. This, of course, is an easy object for the three-inch with a low magnifying power. It would be useless to look for the two fainter companions of delta, discovered by Burnham, even with our five-inch glass. But we shall probably need the five-inch for our next attempt, and it will be well to put on a high power, say three hundred diameters. The star to be examined is the little brilliant dangling below the right-hand end of the Belt, toward Rigel. It appears on the map as eta. Spare no pains in getting an accurate focus, for here is something worth looking at, and unless you have a trained eye you will not easily see it. The star is double, magnitudes third and sixth, and the distance from center to center barely exceeds 1", p. 87? A little tremulousness of the atmosphere for a moment conceals the smaller star, although its presence is manifest from the peculiar jutting of light on one side of the image of the primary. But in an instant the disturbing undulations pass, the air steadies, the image shrinks and sharpens, and two points of piercing brightness, almost touching one another, dart into sight, the more brilliant one being surrounded by an evanescent circle, a tiny ripple of light, which, as it runs round the star and then recedes, alternately embraces and releases the smaller companion. The wash of the light-waves in the atmosphere provokes many expressions of impatience from the astronomer, but it is often a beautiful phenomenon nevertheless.

Between eta and delta is a fifth-magnitude double star, Sigma 725, which is worth a moment's attention. The primary, of a reddish color, has a very faint star, eleventh magnitude, at a distance of 12.7", p. 88?

Still retaining the five-inch in use, we may next turn to the other end of the Belt, where, just under zeta, we perceive the fourth-magnitude star sigma. He must be a person of indifferent mind who, after looking with unassisted eyes at the modest glimmering of this little star, can see it as the telescope reveals it without a thrill of wonder and a cry of pleasure. The glass, as by a touch of magic, changes it from one into eight or ten stars. There are two quadruple sets three and a half minutes of arc apart. The first set exhibits a variety of beautiful colors. The largest star, of fourth magnitude, is pale gray; the second in rank, seventh magnitude, distance 42", p. 61? presents a singular red, "grape-red" Webb calls it; the third, eighth magnitude, distance 12", p. 84? is blue; and the fourth, eleventh magnitude, distance 12", p. 236? is apparently white. Burnham has doubled the fourth-magnitude star, distance 0.23". The second group of four stars consists of three of the eighth to ninth magnitude, arranged in a minute triangle with a much fainter star near them. Between the two quadruple sets careful gazing reveals two other very faint stars. While the five-inch gives a more satisfactory view of this wonderful multiple star than any smaller telescope can do, the four-inch and even the three-inch would have shown it to us as a very beautiful object. However we look at them, there is an appearance of association among these stars, shining with their contrasted colors and their various degrees of brilliance, which is significant of the diversity of conditions and circumstances under which the suns and worlds beyond the solar walk exist.

From sigma let us drop down to see the wonders of Orion's Sword displayed just beneath. We can use with advantage any one of our three telescopes; but since we are going to look at a nebula, it is fortunate that we have a glass so large as five inches aperture. It will reveal interesting things that escape the smaller instruments, because it grasps more than one and a half times as much light as the four-inch, and nearly three times as much as the three-inch; and in dealing with nebulaea plenty of light is the chief thing to be desired. The middle star in the Sword is theta, and is surrounded by the celebrated Nebula of Orion. The telescope shows theta separated into four stars arranged at the corners of an irregular square, and shining in a black gap in the nebula. These four stars are collectively named the Trapezium. The brightest is of the sixth magnitude, the others are of the seventh, seven and a half, and eighth magnitudes respectively. The radiant mist about them has a faint greenish tinge, while the four stars, together with three others at no great distance, which follow a fold of the nebula like a row of buttons on a coat, always appear to me to show an extraordinary liveliness of radiance, as if the strange haze served to set them off.

Our three-inch would have shown the four stars of the Trapezium perfectly well, and the four-inch would have revealed a fifth star, very faint, outside a line joining the smallest of the four and its nearest neighbor. But the five-inch goes a step farther and enables us, with steady gazing to see even a sixth star, of only the twelfth magnitude, just outside the Trapezium, near the brightest member of the quartet. The Lick telescope has disclosed one or two other minute points of light associated with the Trapezium. But more interesting than the Trapezium is the vast cloud, full of strange shapes, surrounding it. Nowhere else in the heavens is the architecture of a nebula so clearly displayed. It is an unfinished temple whose gigantic dimensions, while exalting the imagination, proclaim the omnipotence of its builder. But though unfinished it is not abandoned. The work of creation is proceeding within its precincts. There are stars apparently completed, shining like gems just dropped from the hand of the polisher, and around them are masses, eddies, currents, and swirls of nebulous matter yet to be condensed, compacted, and constructed into suns. It is an education in the nebular theory of the universe merely to look at this spot with a good telescope. If we do not gaze at it long and wistfully, and return to it many times with unflagging interest, we may be certain that there is not the making of an astronomer in us.

Before quitting the Orion nebula do not fail to notice an eighth-magnitude star, a short distance northeast of the Great Nebula, and nearly opposite the broad opening in the latter that leads in toward the gap occupied by the Trapezium. This star is plainly enveloped in nebulosity, that is unquestionably connected with the larger mass of which it appears to form a satellite.

At the lower end of the Sword is the star iota, somewhat under the third magnitude. Our three-inch will show that it has a bluish companion of seventh or eighth magnitude, at a little more than 11" distance, p. 142? and the larger apertures will reveal a third star, of tenth magnitude, and reddish in color, distance 49", p. 103? Close by iota we find the little double star

Sigma 747, whose components are of five and a half and six and a half magnitudes respectively, and separated 36", p. 223? Above the uppermost star in the Sword is a small star cluster, No. 1184, which derives a special interest from the fact that it incloses a delicate double star, Sigma 750, whose larger component is of the sixth magnitude, while the smaller is of the ninth, and the distance is only 4.3", p. 59? We may try the four-inch on this object.

Having looked at alpha (Betelgeuse), the great topaz star on Orion's right shoulder, and admired the splendor of its color, we may turn the four-inch upon the star Sigma 795, frequently referred to by its number as "52 Orionis." It consists of one star of the sixth and another of sixth and a half magnitude, only 1.5" apart, p. 200? Having separated them with a power of two hundred and fifty diameters on the four-inch, we may try them with a high power on the three-inch. We shall only succeed this time if our glass is of first-rate quality and the air is perfectly steady.

The star lambda in Orion's head presents an easy conquest for the three-inch, as it consists of a light-yellow star of magnitude three and a half and a reddish companion of the sixth magnitude; distance 4", p. 43? There is also a twelfth-magnitude star at 27", p. 183? and a tenth or eleventh magnitude one at 149", p. 278? These are tests for the five-inch, and we must not be disappointed if we do not succeed in seeing the smaller one even with that aperture.

Other objects in Orion, to be found with the aid of our map, are: Sigma 627, a double star, magnitude six and a half and seven, distance 21", p. 260? Omicron Sigma 98, otherwise named iota Orionis, double, magnitude six and seven, distance 1", p. 180? requires five-inch glass; Sigma 652, double, magnitudes six and a half and eight, distance 1.7", p. 184? rho, double, magnitudes five and eight and a half, the latter blue, distance 7", p. 62? may be tried with a three-inch; tau, triple star, magnitudes four, ten and a half, and eleven, distances 36", p. 249? and 36", p. 60? Burnham discovered that the ten-and-a-half magnitude star is again double, distance 4", p. 50? There is not much satisfaction in attempting tau Orionis with telescopes of ordinary apertures; Sigma 629 otherwise m Orionis, double, magnitudes five and a half (greenish) and seven, distance 31.7", p. 28? a pretty object; Sigma 728, otherwise A 32, double, magnitudes five and seven, distance, 0.5" or less, p. 206? a rapid binary,[2] which is at present too close for ordinary telescopes, although it was once within their reach; Sigma 729, double, magnitudes six and eight, distance 2", p. 26? the smaller star pale blue--try it with a four-inch, but five-inch is better; Sigma 816, double, magnitudes six and half and eight and a half, distance 4", p. 289? psi 2, double, magnitudes five and a half and eleven, distance 3", or a little less, p. 322? 905, star cluster, contains about twenty stars from the eighth to the eleventh magnitude; 1267, nebula, faint, containing a triple star of the eighth magnitude, two of whose components are 51" apart, while the third is only 1.7" from its companion, p. 85? 1376, star cluster, small and crowded; 1361, star cluster, triangular shape, containing thirty stars, seventh to tenth magnitudes, one of which is a double, distance 2.4".

[2] The term "binary" is used to describe double stars which are in motion about their common center of gravity.

Let us now leave the inviting star-fields of Orion and take a glance at the little constellation of Lepus, crouching at the feet of the mythical giant. We may begin with a new kind of object, the celebrated red variable R Leporis (map No. 1). This star varies from the sixth or seventh magnitude to magnitude eight and a half in a period of four hundred and twenty-four days. Hind's picturesque description of its color has frequently been quoted. He said it is "of the most intense crimson, resembling a blood-drop on the black ground of the sky." It is important to remember that this star is reddest when faintest, so that if we chance to see it near its maximum of brightness it will not impress us as being crimson at all, but rather a dull, coppery red. Its spectrum indicates that it is smothered with absorbing vapors, a sun near extinction which, at intervals, experiences an accession of energy and bursts through its stifling envelope with explosive radiance, only to faint and sink once more. It is well to use our largest aperture in examining this star.

We may also employ the five-inch for an inspection of the double star iota, whose chief component of the fifth magnitude is beautifully tinged with green. The smaller companion is very faint, eleventh magnitude, and the distance is about 13", p. 337?

Another fine double in Lepus is kappa, to be found just below iota; the components are of the fifth and eighth magnitudes, pale yellow and blue respectively, distance 2.5", p. 360? the third-magnitude star alpha has a tenth-magnitude companion at a distance of 35", p. 156? and its neighbor beta (map No. 2), according to Burnham, is attended by three eleventh-magnitude stars, two of which are at distances of 206", p. 75? and 240", p. 58? respectively, while the third is less than 3" from beta, p. 288? the star gamma (map No. 2) is a wide double, the distance being 94", and the magnitudes four and eight. The star numbered 45 is a remarkable multiple, but the components are too faint to possess much interest for those who are not armed with very powerful telescopes.

From Lepus we pass to Canis Major (map No. 2). There is no hope of our being able to see the companion of alpha (Sirius), at present (1901), even with our five-inch. Discovered by Alvan Clark with an eighteen-inch telescope in 1862, when its distance was 10" from the center of Sirius, this ninth-magnitude star has since been swallowed up in the blaze of its great primary. At first, it slightly increased its distance, and from 1868 until 1879 most of the measures made by different observers considerably exceeded 11". Then it began to close up, and in 1890 the distance scarcely exceeded 4". Burnham was the last to catch sight of it with the Lick telescope in that year. After that no human eye saw it until 1896, when it was rediscovered at the Lick Observatory. Since then the distance has gradually increased to nearly 5". According to Burnham, its periodic time is about fifty-three years, and its nearest approach to Sirius should have taken place in the middle of 1892. Later calculations reduce the periodic time to forty-eight or forty-nine years. If we can not see the companion of the Dog Star with our instruments, we can at least, while admiring the splendor of that dazzling orb, reflect with profit upon the fact that although the companion is ten thousand times less bright than Sirius, it is half as massive as its brilliant neighbor. Imagine a subluminous body half as ponderous as the sun to be set revolving round it somewhere between Uranus and Neptune. Remember that that body would possess one hundred and sixty-five thousand times the gravitating energy of the earth, and that five hundred and twenty Jupiters would be required to equal its power of attraction, and then consider the consequences to our easy-going planets! Plainly the solar system is not cut according to the Sirian fashion. We shall hardly find a more remarkable coupling of celestial bodies until we come, on another evening, to a star that began, ages ago, to amaze the thoughtful and inspire the superstitious with dread--the wonderful Algol in Perseus.

We may remark in passing that Sirius is the brightest representative of the great spectroscopic type I, which includes more than half of all the stars yet studied, and which is characterized by a white or bluish-white color, and a spectrum possessing few or at best faint metallic lines, but remarkably broad, black, and intense lines of hydrogen. The inference is that Sirius is surrounded by an enormous atmosphere of hydrogen, and that the intensity of its radiation is greater, surface for surface, than that of the sun. There is historical evidence to support the assertion, improbable in itself, that Sirius, within eighteen hundred years, has changed color from red to white.

With either of our telescopes we shall have a feast for the eye when we turn the glass upon the star cluster No. 1454, some four degrees south of Sirius. Look for a red star near the center. Observe the curving rows so suggestive of design, or rather of the process by which this cluster was evolved out of a pre-existing nebula. You will recall the winding streams in the Great Nebula of Orion. Another star cluster worth a moment's attention is No. 1479, above and to the left of Sirius. We had better use the five-inch for this, as many of the stars are very faint. Not far away we find the double star , whose components are of the fifth and eighth magnitudes, distance 2.8", p. 343? The small star is pale blue. Cluster No. 1512 is a pleasing object with our largest aperture. In No. 1511 we have a faint nebula remarkable for the rows of minute stars in and near it. The star gamma is an irregular variable. In 1670 it is said to have almost disappeared, while at the beginning of the eighteenth century it was more than twice as bright as it is to-day. The reddish star delta is also probably variable. In my "Astronomy with an Opera Glass" will be found a cut showing a singular array of small stars partly encircling delta. These are widely scattered by a telescope, even with the lowest power.

Eastward from Canis Major we find some of the stars of Argo Navis. Sigma 1097, of the sixth magnitude, has two minute companions at 20" distance, p. 311?and 312? The large star is itself double, but the distance, 0.8", p. 166? places it beyond our reach. According to Burnham, there is yet a fourth faint star at 31", p. 40? Some three degrees and a half below and to the left of the star just examined is a beautiful star cluster, No. 1551. Nos. 1564, 1571, and 1630 are other star clusters well worth examination. A planetary nebula is included in 1564. With very powerful telescopes this nebula has been seen ring-shaped. Sigma 1146, otherwise known as 5 Navis, is a pretty double, colors pale yellow and blue, magnitudes five and seven, distance 3.25", p. 19?

Our three-inch will suffice for this.

North of Canis Major and Argo we find Monoceros and Canis Minor (map No. 3). The stars forming the western end of Monoceros are depicted on map No. 1. We shall begin with these. The most interesting and beautiful is 11, a fine triple star, magnitudes five, six, and seven, distances 7.4", p. 131? and 2.7", p. 103? Sir William Herschel regarded this as one of the most beautiful sights in the heavens. It is a good object to try our three-inch on, although it should not be difficult for such an aperture. The star 4 is also a triple, magnitudes six, ten, and eleven, distances 3.4", p. 178? and 10", p. 244? We should glance at the star 5 to admire its fine orange color. In 8 we find a golden fifth-magnitude star, combined with a blue or lilac star of the seventh magnitude, distance 14", p. 24? Sigma 938 is a difficult double, magnitudes six and a half and twelve, distance 10", p. 210? Sigma 921 is double, magnitudes six and a half and eight, distance 16", p. 4? At the spot marked on the map 1424 we find an interesting cluster containing one star of the sixth magnitude.

The remaining stars of Monoceros will be found on map No. 3. The double and triple stars to be noted are S, or Sigma 950 (which is also a variable and involved in a faint nebula), magnitudes six and nine, distance 2.5", p. 206? Sigma 1183, double, magnitudes five and a half and eight, distance 31", p. 326? Sigma 1190, triple, magnitudes five and a half, ten, and nine, distances 31", p. 105? and 67", p. 244? The clusters are 1465, which has a minute triple star near the center; 1483, one member of whose swarm is red; 1611, very small but rich; and 1637, interesting for the great number of ninth-magnitude stars that it contains. We should use the five-inch for all of these.

Canis Minor and the Head of Hydra are also contained on map No. 3. Procyon, alpha of Canis Minor, has several minute stars in the same field of view. There is, besides, a companion which, although it was known to exist, no telescope was able to detect until November, 1896. It must be of immense mass, since its attraction causes perceptible perturbations in the motion of Procyon. Its magnitude is eight and a half, distance 4.83", p. 338? One of the small stars just referred to, the second one east of Procyon, distant one third of the moon's diameter, is an interesting double. Our four-inch may separate it, and the five-inch is certain to do so. The magnitudes are seven and seven and a half or eight, distance 1.2", p. 133? This star is variously named Sigma 1126 and 31 Can. Min. Bode. Star No. 14 is a wide triple, magnitudes six, seven, and eight, distances 75, p. 65? and 115", p. 154?

PROCYON AND ITS NEIGHBORS.

In the Head of Hydra we find Sigma 1245, a double of the sixth and seventh magnitudes, distance 10.5", p. 25? The larger star shows a fine yellow. In epsilon we have a beautiful combination of a yellow with a blue star, magnitudes four and eight, distance 3.4", p. 198? Finally, let us look at theta for a light test with the five-inch. The two stars composing it are of the fourth and twelfth magnitudes, distance 50", p. 170?

The brilliant constellations of Gemini and Taurus tempt us next, but warning clouds are gathering, and we shall do well to house our telescopes and warm our fingers by the winter fire. There will be other bright nights, and the stars are lasting.

CHAPTER III

FROM GEMINI TO LEO AND ROUND ABOUT

"If thou wouldst gaze on starry Charioteer, And hast heard legends of the wondrous Goat, Vast looming shalt thou find on the Twins' left, His form bowed forward."--POSTE'S ARATUS.

The zodiacal constellations of Gemini, Cancer, and Leo, together with their neighbors Auriga, the Lynx, Hydra, Sextans, and Coma Berenices, will furnish an abundance of occupation for our second night at the telescope. We shall begin, using our three-inch glass, with alpha, the chief star of Gemini (map No. 4). This is ordinarily known as Castor. Even an inexperienced eye perceives at once that it is not as bright as its neighbor Pollux, beta. Whether this fact is to be regarded as indicating that Castor was brighter than Pollux in 1603, when Bayer attached their Greek letters, is still an unsettled question. Castor may or may not be a variable, but it is, at any rate, one of the most beautiful double stars in the heavens. A power of one hundred is amply sufficient to separate its components, whose magnitudes are about two and three, the distance between them being 6", p. 226? A slight yet distinct tinge of green, recalling that of the Orion nebula, gives a peculiar appearance to this couple. Green is one of the rarest colors among the stars. Castor belongs to the same general spectroscopic type in which Sirius is found, but its lines of hydrogen are broader than those seen in the spectrum of the Dog Star. There is reason for thinking that it may be surrounded with a more extensive atmosphere of that gaseous metal called hydrogen than any other bright star possesses. There seems to be no doubt that the components of Castor are in revolution around their common center of gravity, although the period is uncertain, varying in different estimates all the way from two hundred and fifty to one thousand years; the longer estimate is probably not far from the truth. There is a tenth-magnitude star, distance 73", p. 164? which may belong to the same system.

From Castor let us turn to Pollux, at the same time exchanging our three-inch telescope for the four-inch, or, still better, the five-inch. Pollux has five faint companions, of which we may expect to see three, as follows: Tenth magnitude, distance 175", p. 70? nine and a half magnitude, distance 206", p. 90? and ninth magnitude, distance 229", p. 75? Burnham has seen a star of thirteen and a half magnitude, distance 43", p. 275? and has divided the tenth-magnitude star into two components, only 1.4" apart, the smaller being of the thirteenth magnitude, and situated at the angle 128? A calculation based on Dr. Elkin's parallax of 0.068" for Pollux shows that that star may be a hundredfold more luminous than the sun, while its nearest companion may be a body smaller than our planet Jupiter, but shining, of course, by its own light. Its distance from Pollux, however, exceeds that of Jupiter from the sun in the ratio of about one hundred and thirty to one.

In the double star pi we shall find a good light test for our three-inch aperture, the magnitudes being six and eleven, distance 22", p. 212? The four-inch will show that kappa is a double, magnitudes four and ten, distance 6", p. 232? The smaller star is of a delicate blue color, and it has been suspected of variability. That it may be variable is rendered the more probable by the fact that in the immediate neighborhood of kappa there are three undoubted variables, S, T, and U, and there appears to be some mysterious law of association which causes such stars to group themselves in certain regions. None of the variables just named ever become visible to the naked eye, although they all undergo great changes of brightness, sinking from the eighth or ninth magnitude down to the thirteenth or even lower. The variable R, which lies considerably farther west, is well worth attention because of the remarkable change of color which it sometimes exhibits. It has been seen blue, red, and yellow in succession. It varies from between the sixth and seventh magnitudes to less than the thirteenth in a period of about two hundred and forty-two days.

Not far away we find a still more curious variable zeta; this is also an interesting triple star, its principal component being a little under the third magnitude, while one of the companions is of the seventh magnitude, distance 90", p. 355? and the other is of the eleventh magnitude or less, distance 65", p. 85? We should hardly expect to see the fainter companion with the three-inch. The principal star varies from magnitude three and seven tenths down to magnitude four and a half in a period of a little more than ten days.

With the four-or five-inch we get a very pretty sight in delta, which appears split into a yellow and a purple star, magnitudes three and eight, distance 7", p. 206?

Near delta, toward the east, lies one of the strangest of all the nebulae (See the figures 1532 on the map.) Our telescopes will show it to us only as a minute star surrounded with a nebulous atmosphere, but its appearance with instruments of the first magnitude is so astonishing and at the same time so beautiful that I can not refrain from giving a brief description of it as I saw it in 1893 with the great Lick telescope. In the center glittered the star, and spread evenly around it was a circular nebulous disk, pale yet sparkling and conspicuous. This disk was sharply bordered by a narrow black ring, and outside the ring the luminous haze of the nebula again appeared, gradually fading toward the edge to invisibility. The accompanying cut, which exaggerates the brightness of the nebula as compared with the star, gives but a faint idea of this most singular object. If its peculiarities were within the reach of ordinary telescopes, there are few scenes in the heavens that would be deemed equally admirable.

In the star eta we have another long-period variable, which is also a double star; unfortunately the companion, being of only the tenth magnitude and distant less than 1" from its third-magnitude primary, is beyond the reach of our telescopes. But eta points the way to one of the finest star clusters in the sky, marked 1360 on the map. The naked eye perceives that there is something remarkable in that place, and the opera glass faintly reveals its distant splendors, but the telescope fairly carries us into its presence. Its stars are innumerable, varying from the ninth magnitude downward to the last limit of visibility, and presenting a wonderful array of curves which are highly interesting from the point of view of the nebular origin of such clusters. Looking backward in time, with that theory to guide us, we can see spiral lines of nebulous mist occupying the space that now glitters with interlacing rows of stars. It is certainly difficult to understand how such lines of nebula could become knotted with the nuclei of future stars, and then gradually be absorbed into those stars; and yet, if such a process does not occur, what is the meaning of that narrow nebulous streak in the Pleiades along which five or six stars are strung like beads on a string? The surroundings of this cluster, 1360, as one sweeps over them with the telescope gradually drawing toward the nucleus, have often reminded me of the approaches to such a city as London. Thicker and closer the twinkling points become, until at last, as the observers eye follows the gorgeous lines of stars trending inward, he seems to be entering the streets of a brilliantly lighted metropolis.

Other objects in Gemini that we can ill miss are: , double, magnitudes three and eleven, distance 73", p. 76? colors yellow and blue; 15, double, magnitudes six and eight, distance 33", p. 205? gamma, remarkable for array of small stars near it; 38, double, magnitudes six and eight, distance 6.5", p. 162? colors yellow and blue (very pretty); lambda, double, magnitudes four and eleven, distance 10", p. 30? color of larger star blue--try with the five-inch; epsilon, double, magnitudes three and nine, distance 110", p. 94?

From Gemini we pass to Cancer. This constellation has no large stars, but its great cluster is easily seen as a starry cloud with the naked eye. With the telescope it presents the most brilliant appearance with a very low power. It was one of the first objects that Galileo turned to when he had completed his telescope, and he wonderingly counted its stars, of which he enumerated thirty-six, and made a diagram showing their positions.

The most interesting star in Cancer is zeta, a celebrated triple. The magnitudes of its components are six, seven, and seven and a half; distances 1.14", p. 6? and 5.7", p. 114? We must use our five-inch glass in order satisfactorily to separate the two nearest stars. The gravitational relationship of the three stars is very peculiar. The nearest pair revolve around their common center in about fifty-eight years, while the third star revolves with the other two, around a center common to all three, in a period of six or seven hundred years. But the movements of the third star are erratic, and inexplicable except upon the hypothesis advanced by Seeliger, that there is an invisible, or dark, star near it by whose attraction its motion is perturbed.

In endeavoring to picture the condition of things in zeta Cancri we might imagine our sun to have a companion sun, a half or a third as large as itself, and situated within what may be called planetary distance, circling with it around their center of gravity; while a third sun, smaller than the second and several times as far away, and accompanied by a black or non-luminous orb, swings with the first two around another center of motion. There you would have an entertaining complication for the inhabitants of a system of planets!

Other objects in Cancer are: Sigma 1223, double star, magnitudes six and six and a half, distance 5", p. 214? Sigma 1291, double, magnitudes both six, distance 1.3", p. 328?-four-inch should split it; iota, double, magnitudes four and a half and six and a half, distance 30", p. 308? 66, double magnitudes six and nine, distance 4.8", p. 136? Sigma 1311, double, magnitudes both about the seventh, distance 7", p. 200? 1712, star cluster, very beautiful with the five-inch glass.

[Illustration: MAP NO. 5.]

The constellation of Auriga may next command our attention (map No. 5). The calm beauty of its leading star Capella awakens an admiration that is not diminished by the rivalry of Orion's brilliants glittering to the south of it. Although Capella must be an enormously greater sun than ours, its spectrum bears so much resemblance to the solar spectrum that a further likeness of condition is suggested. No close telescopic companion to Capella has been discovered. A ninth-magnitude companion, distant 159", p. 146? and two others, one of twelfth magnitude at 78", p. 317? the other of thirteenth magnitude at 126", p. 183? may be distant satellites of the great star, but not planets in the ordinary sense, since it is evident that they are self-luminous. It is a significant fact that most of the first-magnitude stars have faint companions which are not so distant as altogether to preclude the idea of physical relationship.

But while Capella has no visible companion, Campbell, of the Lick

Observatory, has lately discovered that it is a conspicuous example of a peculiar class of binary stars only detected within the closing decade of the nineteenth century. The nature of these stars, called spectroscopic binaries, may perhaps best be described while we turn our attention from Capella to the second star in Auriga beta (Menkalina), which not only belongs to the same class, but was the first to be discovered. Neither our telescopes, nor any telescope in existence, can directly reveal the duplicity of beta Aurig to the eye--i. e., we can not see the two stars composing it, because they are so close that their light remains inextricably mingled after the highest practicable magnifying power has been applied in the effort to separate them. But the spectroscope shows that the star is double and that its components are in rapid revolution around one another, completing their orbital swing in the astonishingly short period of four days! The combined mass of the two stars is estimated to be two and a half times the mass of the sun, and the distance between them, from center to center, is about eight million miles.

The manner in which the spectroscope revealed the existence of two stars in beta Aurig?is a beautiful illustration of the unexpected and, so to speak, automatic application of an old principle in the discovery of new facts not looked for. It was noticed at the Harvard Observatory that the lines in the photographed spectrum of beta Aurig?(and of a few other stars to be mentioned later) appeared single in some of the photographs and double in others. Investigation proved that the lines were doubled at regular intervals of about two days, and that they appeared single in the interim. The explanation was not far to seek. It is known that all stars which are approaching us have their spectral lines shifted, by virtue of their motion of approach, toward the violet end of the spectrum, and that, for a similar reason, all stars which are receding have their lines shifted toward the red end of the spectrum. Now, suppose two stars to be revolving around one another in a plane horizontal, or nearly so, to the line of sight. When they are at their greatest angular distance apart as seen from the earth one of them will evidently be approaching at the same moment that the other is receding. The spectral lines of the first will therefore be shifted toward the violet, and those of the second will be shifted toward the red. Then if the stars, when at their greatest distance apart, are still so close that the telescope can not separate them, their light will be combined in the spectrum; but the spectral lines, being simultaneously shifted in opposite directions, will necessarily appear to be doubled. As the revolution of the stars continues, however, it is clear that their motion will soon cease to be performed in the line of sight, and will become more and more athwart that line, and as this occurs the spectral lines will gradually assume their normal position and appear single. This is the sequence of phenomena in beta Aurig? And the same sequence is found in Capella and in several other more or less conspicuous stars in various parts of the heavens.

Such facts, like those connecting rows and groups of stars with masses and spiral lines of nebula are obscure signboards, indicating the opening of a way which, starting in an unexpected direction, leads deep into the mysteries of the universe.

Southward from beta we find the star theta, which is a beautiful quadruple. We shall do best with our five-inch here, although in a fine condition of the atmosphere the four-inch might suffice. The primary is of the third magnitude; the first companion is of magnitude seven and a half, distance 2", p. 5? the second, of the tenth magnitude, distance 45", p. 292? and the third, of the tenth magnitude, distance 125", p. 350?

We should look at the double Sigma 616 with one of our larger apertures in order to determine for ourselves what the colors of the components are. There is considerable diversity of opinion on this point. Some say the larger star is pale red and the smaller light blue; others consider the color of the larger star to be greenish, and some have even called it white. The magnitudes are five and nine, distance 6", p. 350?

Auriga contains several noteworthy clusters which will be found on the map. The most beautiful of these is 1295, in which about five hundred stars have been counted.

The position of the new star of 1892, known as Nova Aurig is also indicated on the map. While this never made a brilliant appearance, it gave rise to a greater variety of speculative theories than any previous phenomenon of the kind. Although not recognized until January 24, 1892, this star, as photographic records prove, was in existence on December 9, 1891. At its brightest it barely exceeded magnitude four and a half, and its maximum occurred within ten days after its first recognition. When discovered it was of the fifth magnitude. It was last seen in its original form with the Lick telescope on April 26th, when it had sunk to the lowest limit of visibility. To everybody's astonishment it reappeared in the following August, and on the 17th of that month was seen shining with the light of a tenth-magnitude star, but presenting the spectrum of a nebula! Its visual appearance in the great telescope was now also that of a planetary nebula. Its spectrum during the first period of its visibility had been carefully studied, so that the means existed for making a spectroscopic comparison of the phenomenon in its two phases. During the first period, when only a stellar spectrum was noticed, remarkable shiftings of the spectral lines occurred, indicating that two and perhaps three bodies were concerned in the production of the light of the new star, one of which was approaching the earth, while the other or the others receded with velocities of several hundred miles per second! On the revival in the form of a planetary nebula, while the character of the spectrum had entirely changed, evidences of rapid motion in the line of sight remained.

But what was the meaning of all this? Evidently a catastrophe of some kind had occurred out there in space. The idea of a collision involving the transformation of the energy of motion into that of light and heat suggests itself at once. But what were the circumstances of the collision? Did an extinguished sun, flying blindly through space, plunge into a vast cloud of meteoric particles, and, under the lashing impact of so many myriads of missiles, break into superficial incandescence, while the cosmical wrack through which it had driven remained glowing with nebulous luminosity? Such an explanation has been offered by Seeliger. Or was Vogel right when he suggested that Nova Aurig could be accounted for by supposing that a wandering dark body had run into collision with a system of planets surrounding a decrepit sun (and therefore it is to be hoped uninhabited), and that those planets had been reduced to vapor and sent spinning by the encounter, the second outburst of light being caused by an outlying planet of the system falling a prey to the vagabond destroyer? Or some may prefer the explanation, based on a theory of Wilsing's, that two great bodies, partially or wholly opaque and non-luminous at their surfaces, but liquid hot within, approached one another so closely that the tremendous strain of their tidal attraction burst their shells asunder so that their bowels of fire gushed briefly visible, amid a blaze of spouting vapors. And yet Lockyer thinks that there was no solid or semisolid mass concerned in the phenomenon at all, but that what occurred was simply the clash of two immense swarms of meteors that had crossed one another's track.

Well, where nobody positively knows, everybody has free choice. In the meantime, look at the spot in the sky where that little star made its appearance and underwent its marvelous transformation, for, even if you can see no remains of it there, you will feel your interest in the problem it has presented, and in the whole subject of astronomy, greatly heightened and vivified, as the visitor to the field of Waterloo becomes a lover of history on the spot.

The remaining objects of special interest in Auriga may be briefly mentioned: 26, triple star, magnitudes five, eight, and eleven, distances 12", p. 268? and 26", p. 113? 14, triple star, magnitudes five, seven and a half, and eleven, distances 14", p. 224? and 12.6", p. 342? the last difficult for moderate apertures; lambda, double, magnitudes five and nine, distance 121", p. 13? epsilon, variable, generally of third magnitude, but has been seen of only four and a half magnitude; 41, double, magnitudes five and six, distance 8", p. 354? 996, 1067, 1119, and 1166, clusters all well worth inspection, 1119 being especially beautiful.

The inconspicuous Lynx furnishes some fine telescopic objects, all grouped near the northwestern corner of the constellation. Without a six-inch telescope it would be a waste of time to attack the double star 4, whose components are of sixth and eighth magnitudes, distance 0.8", p. 103? but its neighbor, 5, a fine triple, is within our reach, the magnitudes being six, ten, and eight, distances 30", p. 139? and 96", p. 272? In 12 Lyncis we find one of the most attractive of triple stars, which in good seeing weather is not beyond the powers of a three-inch glass, although we shall have a far more satisfactory view of it with the four-inch. The components are of the sixth, seventh, and eighth magnitudes, distances 1.4", p. 117? and 8.7", p. 304? A magnifying power which just suffices clearly to separate the disks of the two nearer stars makes this a fine sight. A beautiful contrast of colors belongs to the double star 14, but unfortunately the star is at present very close, the distance between its sixth and seventh magnitude components not exceeding 0.8", position angle 64? Sigma 958 is a pretty double, both stars being of the sixth magnitude, distance 5", p. 257? Still finer is Sigma 1009, a double, whose stars are both a little above the seventh magnitude and nearly equal, distance 3", p. 156? A low power suffices to show the three stars in 19, their magnitudes being six and a half, seven and a half, and eight, distances 15", p.

312? and 215", p. 358? Webb describes the two smaller stars as plum-colored. Plum-colored suns!

At the opposite end of the constellation are two fine doubles, Sigma 1333, magnitudes six and a half and seven, distance 1.4", p. 39? and 38, magnitudes four and seven, distance 2.9", p. 235?

Under the guidance of map No. 6 we turn to Leo, which contains one of the leading gems among the double stars, gamma, whose components, of the second and fourth magnitudes, are respectively yellow and green, the green star, according to some observers, having a peculiar tinge of red. Their distance apart is 3.7", p. 118? and they are undoubtedly in revolution about a common center, the probable period being about four hundred years. The three-inch glass should separate them easily when the air is steady, and a pleasing sight they are.

The star iota is a closer double, and also very pretty, magnitudes four and eight, colors lemon and light blue, distance 2.17", p. 53? Other doubles are tau, magnitudes five and seven, distance 95", p. 170? 88, magnitudes seven and nine, distance 15", p. 320? 90, triple, magnitudes six, seven and a half, and ten, distance, 3.5", p. 209? and 59", p. 234? 54, magnitudes four and a half and seven, distance 6.2", p. 102? and 49, magnitudes six and nine, distance 2.4", p. 158?

Leo contains a remarkable variable star, R, deep red in color, and varying in a space of a hundred and forty-four days from the fifth to the tenth magnitude. It has also several nebulae of which only one needs special mention, No. 1861. This is spindle-shaped, and large telescopes show that it consists of three nebulae The observer with ordinary instruments finds little to see and little to interest him in these small, faint nebulae

We may just glance at two double stars in the small constellation of Sextans, situated under Leo. These are: 9, magnitudes seven and eight, distance 53", p. 292? and 35, magnitudes six and seven, distance 6.9", p. 240?

Coma Berenices (map No. 6) includes several interesting objects. Let us begin with the star 2, a double, of magnitudes six and seven and a half, distance 3.6", p. 240? The color of the smaller star is lilac. This hue, although not extremely uncommon among double stars elsewhere, recurs again and again, with singular persistence, in this little constellation. For instance, in the very next star that we look at, 12, we find a double whose smaller component is lilac. The magnitudes in 12 are five and eight, distance 66", p. 168? So also the wide double 17, magnitudes five and a half and six, distance 145", exhibits a tinge of lilac in the smaller component; the triple 35, magnitudes five, eight, and nine, distances 1", p. 77? and 28.7", p. 124? has four colors yellow, lilac, and blue, and the double 24, magnitudes five and six, distance 20", p. 270? combines an orange with a lilac star, a very striking and beautiful contrast. We should make a mistake if we regarded this wonderful distribution of color among the double stars as accidental. It is manifestly expressive of their physical condition, although we can not yet decipher its exact meaning.

The binary 42 Com Berenicis is too close for ordinary telescopes, but it is highly interesting as an intermediate between those pairs which the telescope is able to separate and those--like beta Aurig-which no magnifying power can divide, but which reveal the fact that they are double by the periodical splitting of their spectral lines. The orbit in 42 Com Berenicis is a very small one, so that even when the components are at their greatest distance apart they can not be separated by a five-or six-inch glass. Burnham, using the Lick telescope, in 1890 made the distance 0.7"; Hall, using the Washington telescope, in 1891 made it a trifle more than 0.5". He had measured it in 1886 as only 0.27". The period of revolution is believed to be about twenty-five years.

In Coma Berenices there is an outlying field of the marvelous nebulous region of Virgo, which we may explore on some future evening. But the nebulae in Coma are very faint, and, for an amateur, hardly worth the trouble required to pick them up. The two clusters included in the map, 2752 and 3453, are bright enough to repay inspection with our largest aperture.

Although Hydra is the largest constellation in the heavens, extending about seven hours, or 105? in right ascension, it contains comparatively few objects of interest, and most of these are in the head or western end of the constellation, which we examined during our first night at the telescope. In the central portion of Hydra, represented on map No. 7, we find its leading star alpha, sometimes called Alphard, or Cor Hydr? a bright second-magnitude star that has been suspected of variability. It has a decided orange tint, and is accompanied, at a distance of 281", p. 153? by a greenish tenth-magnitude star. Bu. 339 is a fine double, magnitudes eight and nine and a half, distance 1.3", p. 216? The planetary nebula 2102 is about 1' in diameter, pale blue in color, and worth looking at, because it is brighter than most objects of its class. Tempel and Secchi have given wonderful descriptions of it, both finding multitudes of stars intermingled with nebulous matter.

For a last glimpse at celestial splendors for the night, let us turn to the rich cluster 1630, in Argo, just above the place where the stream of the Milky Way--here bright in mid-channel and shallowing toward the shores--separates into two or three currents before disappearing behind the horizon. It is by no means as brilliant as some of the star clusters we have seen, but it gains in beauty and impressiveness from the presence of one bright star that seems to captain a host of inferior luminaries.

CHAPTER IV

VIRGO AND HER NEIGHBORS

Following the order of right ascension, we come next to the little constellations Crater and Corvus, which may be described as standing on the curves of Hydra (map No. 8). Beginning with Crater, let us look first at alpha, a yellow fourth-magnitude star, near which is a celebrated red variable R. With a low power we can see both alpha and R in the same field of view, like a very wide double. There is a third star of ninth magnitude, and bluish in color, near R on the side toward alpha. R is variable both in color and light. When reddest, it has been described as "scarlet," "crimson," and "blood-colored"; when palest, it is a deep orange-red. Its light variation has a period the precise length of which is not yet known. The cycle of change is included between the eighth and ninth magnitudes.

While our three-inch telescope suffices to show R, it is better to use the five-inch, because of the faintness of the star. When the color is well seen, the contrast with alpha is very pleasing.

There is hardly anything else in Crater to interest us, and we pass over the border into Corvus, and go at once to its chief attraction, the star delta. The components of this beautiful double are of magnitudes three and eight; distance 24", p. 211? colors yellow and purple.

The night being dark and clear, we take the five-inch and turn it on the nebula 3128, which the map shows just under the border of Corvus in the edge of Hydra. Herschel believed he had resolved this into stars. It is a faint object and small, not exceeding one eighth of the moon's diameter.

Farther east in Hydra, as indicated near the left-hand edge of map No. 8, is a somewhat remarkable variable, R Hydr? This star occasionally reaches magnitude three and a half, while at minimum it is not much above the tenth magnitude. Its period is about four hundred and twenty-five days.

While we have been examining these comparatively barren regions, glad to find one or two colored doubles to relieve the monotony of the search, a glittering white star has frequently drawn our eyes eastward and upward. It is Spica, the great gem of Virgo, and, yielding to its attraction, we now enter the richer constellation over which it presides (map No. 9). Except for its beauty, which every one must admire, Spica, or alpha Virginis, has no special claim upon our attention. Some evidence has been obtained that, like beta Aurig?and Capella, it revolves with an invisible companion of great mass in an orbit only six million miles in diameter. Spica's spectrum resembles that of Sirius. The faint star which our larger apertures show about 6' northeast of Spica is of the tenth magnitude.

Sweeping westward, we come upon Sigma 1669, a pretty little double with nearly equal components of about the sixth magnitude, distance 5.6", p. 124? But our interest is not fully aroused until we reach gamma, a star with a history. The components of this celebrated binary are both nearly of the third magnitude, distance about 5.8", p. 150? They revolve around their common center in something less than two hundred years. According to some authorities, the period is one hundred and seventy years, but it is not yet certainly ascertained. It was noticed about the beginning of the seventeenth century that gamma Virginis was double. In 1836 the stars were so close together that no telescope then in existence was able to separate them, although it is said that the disk into which they had merged was elongated at Pulkowa. In a few years they became easily separable once more. If the one-hundred-and-seventy-year period is correct, they should continue to get farther apart until about 1921. According to Asaph Hall, their greatest apparent distance is 6.3", and their least apparent distance 0.5"; consequently, they will never again close up beyond the separating power of existing telescopes.

There is a great charm in watching this pair of stars even with a three-inch telescope--not so much on account of what is seen, although they are very beautiful, as on account of what we know they are doing. It is no slight thing to behold two distant stars obeying the law that makes a stone fall to the ground and compels the earth to swing round the sun.

In theta we discover a fine triple, magnitudes four and a half, nine, and ten; distances 7", p. 345? and 65", p. 295? The ninth-magnitude star has been described as "violet," but such designations of color are often misleading when the star is very faint. On the other hand it should not be assumed that a certain color does not exist because the observer can not perceive it, for experience shows that there is a wide difference among observers in the power of the eye to distinguish color. I have known persons who could not perceive the difference of hue in some of the most beautifully contrasted colored doubles to be found in the sky. I am acquainted with an astronomer of long experience in the use of telescopes, whose eye is so deficient in color sense that he denies that there are any decided colors among the stars. Such persons miss one of the finest pleasures of the telescope. In examining theta Virginis we shall do best to use our largest aperture, viz., the five-inch. Yet Webb records that all three of the stars in this triple have been seen with a telescope of only three inches aperture. The amateur must remember in such cases how much depends upon practice as well as upon the condition of the atmosphere. There are lamentably few nights in a year when even the best telescope is ideally perfect in performance, but every night's observation increases the capacity of the eye, begetting a kind of critical judgment which renders it to some extent independent of atmospheric vagaries. It will also be found that the idiosyncrasies of the observer are reflected in his instrument, which seems to have its fits of excellence, its inspirations so to speak, while at other times it behaves as if all its wonderful powers had departed.

Another double that perhaps we had better not try with less than four inches aperture is 84 Virginis. The magnitudes are six and nine; distance, 3.5", p. 233? Colors yellow and blue. Sigma 1846 is a fifth-magnitude star with a tenth-magnitude companion, distance only 4", p. 108? Use the five-inch.

And now we approach something that is truly marvelous, the "Field of the Nebulae" This strange region, lying mostly in the constellation Virgo, is roughly outlined by the stars beta, eta, gamma, delta, and epsilon, which form two sides of a square some 15 across. It extends, however, for some distance into Coma Berenices, while outlying nebulae belonging to it are also to be found in the eastern part of Leo. Unfortunately for those who expect only brilliant revelations when they look through a telescope, this throng of nebulae consists of small and inconspicuous wisps as ill defined as bits of thistle-down floating high in the air. There are more than three hundred of them all told, but even the brightest are faint objects when seen with the largest of our telescopes. Why do they congregate thus? That is the question which lends an interest to the assemblage that no individual member of it could alone command. It is a mystery, but beyond question it is explicable. The explanation, however, is yet to be discovered.

The places of only three of the nebulae are indicated on the map. No. 2806 has been described as resembling in shape a shuttle. Its length is nearly one third of the moon's diameter. It is brightest near the center, and has several faint companions. No. 2961 is round, 4' in diameter, and is accompanied by another round nebula in the same field of view toward the south. No. 3105 is double, and powerful telescopes show two more ghostly companions. There is an opportunity for good and useful work in a careful study of the little nebulae that swim into view all over this part of Virgo. Celestial photography has triumphs in store for itself here.

Scattered over and around the region where the nebulae are thickest we find eight or nine variable stars, three of the most remarkable of which, R, S, and U, may be found on the map. R is very irregular, sometimes attaining magnitude six and a half, while at other times its maximum brightness does not exceed that of an eighth-magnitude star. At minimum it sinks to the tenth or eleventh magnitude. Its period is one hundred and forty-five days. U varies from magnitude seven or eight down to magnitude twelve or under and then regains its light, in a period of about two hundred and seven days. S is interesting for its brilliant red color. When brightest, it exceeds the sixth magnitude, but at some of its maxima the magnitude is hardly greater than the eighth. At minimum it goes below the twelfth magnitude. Period, three hundred and seventy-six days.

Next east of Virgo is Libra, which contains a few notable objects (map No. 10). The star alpha has a fifth-magnitude companion, distant about 230", which can be easily seen with an opera glass. At the point marked A on the map is a curious multiple star, sometimes referred to by its number in Piazzi's catalogues as follows: 212 P. xiv. The two principal stars are easily seen, their magnitudes being six and seven and a half; distance 15", p. 290? Burnham found four other faint companions, for which it would be useless for us to look. The remarkable thing is that these faint stars, the nearest of which is distant about 50" from the largest member of the group and the farthest about 129", do not share, according to their discoverer, in the rapid proper motion of the two main stars.

In iota we find a double a little difficult for our three-inch. The components are of magnitudes four and a half and nine, distance 57", p. 110? Burnham discovered that the ninth-magnitude star consists of two of the tenth less than 2" apart, p. 24?

No astronomer who happens to be engaged in this part of the sky ever fails, unless his attention is absorbed by something of special interest, to glance at beta Libr? which is famous as the only naked-eye star having a decided green color. The hue is pale, but manifest.[3]

[3] Is the slight green tint perceptible in Sirius variable? I am sometimes disposed to think it is.

The star is a remarkable variable, belonging to what is called the Algol type. Its period, according to Chandler, is 2 days 7 hours, 51 minutes, 22.8 seconds. The time occupied by the actual changes is about twelve hours. At maximum the star is of magnitude five and at minimum of magnitude 6.2.

We may now conveniently turn northward from Virgo in order to explore one of the most interesting of the constellations (map No. 11). Its leading star alpha, Arcturus, is the brightest in the northern hemisphere. Its precedence over its rivals Vega and Capella, long in dispute, has been settled by the Harvard photometry. You notice that the color of Arcturus, when it has not risen far above the horizon, is a yellowish red, but when the star is near midheaven the color fades to light yellow. The hue is possibly variable, for it is recorded that in 1852 Arcturus appeared to have nearly lost its color. If it should eventually turn white, the fact would have an important bearing upon the question whether Sirius was, as alleged, once a red or flame-colored star.

But let us sit here in the starlight, for the night is balmy, and talk about Arcturus, which is perhaps actually the greatest sun within the range of terrestrial vision. Its parallax is so minute that the consideration of the tremendous size of this star is a thing that the imagination can not placidly approach. Calculations, based on its assumed distance, which show that it outshines the sun several thousand times, may be no exaggeration of the truth! It is easy to make such a calculation. One of Dr. Elkin's parallaxes for Arcturus is 0.018". That is to say, the displacement of Arcturus due to the change in the observer's point of view when he looks at the star first from one side and then from the other side of the earth's orbit, 186,000,000 miles across, amounts to only eighteen one-thousandths of a second of arc. We can appreciate how small that is when we reflect that it is about equal to the apparent distance between the heads of two pins placed an inch apart and viewed from a distance of a hundred and eighty miles!

Assuming this estimate of the parallax of Arcturus, let us see how it will enable us to calculate the probable size or light-giving power of the star as compared with the sun. The first thing to do is to multiply the earth's distance from the sun, which may be taken at 93,000,000 miles, by 206,265, the number of seconds of arc in a radian, the base of circular measure, and then divide the product by the parallax of the star. Performing the multiplication and division, we get the following:

19,182,645,000,000 / .018 = 1,065,702,500,000,000.

The quotient represents miles! Call it, in round numbers, a thousand millions of millions of miles. This is about 11,400,000 times the distance from the earth to the sun.

Now for the second part of the calculation: The amount of light received on the earth from some of the brighter stars has been experimentally compared with the amount received from the sun. The results differ rather widely, but in the case of Arcturus the ratio of the star's light to sunlight may be taken as about one twenty-five-thousand-millionth--i. e., 25,000,000,000 stars, each equal to Arcturus, would together shed upon the earth as much light as the sun does. But we know that light varies inversely as the square of the distance; for instance, if the sun were twice as far away as it is, its light would be diminished for us to a quarter of its present amount. Suppose, then, that we could remove the earth to a point midway between the sun and Arcturus, we should then be 5,700,000 times as far from the sun as we now are. In order to estimate how much light the sun would send us from that distance we must square the number 5,700,000 and then take the result inversely, or as a fraction. We thus get 1 / 32,490,000,000,000, representing the ratio of the sun's light at half the distance of Arcturus to that at its real distance. But while receding from the sun we should be approaching Arcturus. We should get, in fact, twice as near to that star as we were before, and therefore its light would be increased for us fourfold. Now, if the amount of sunlight had not changed, it would exceed the light of Arcturus only a quarter as much as it did before, or in the ratio of 25,000,000,000 / 4 = 6,250,000,000 to 1. But, as we have seen, the sunlight would diminish through increase of distance to one 32,490,000,000,000th part of its original amount. Hence its altered ratio to the light of Arcturus would become 6,250,000,000 to 32,490,000,000,000, or 1 to 5,198.

This means that if the earth were situated midway between the sun and Arcturus, it would receive 5,198 times as much light from that star as it would from the sun! It is quite probable, moreover, that the heat of Arcturus exceeds the solar heat in the same ratio, for the spectroscope shows that although Arcturus is surrounded with a cloak of metallic vapors proportionately far more extensive than the sun's, yet, smothered as the great star seems in some respects to be, it rivals Sirius itself in the intensity of its radiant energy.

If we suppose the radiation of Arcturus to be the same per unit of surface as the sun's, it follows that Arcturus exceeds the sun about 375,000 times in volume, and that its diameter is no less than 62,350,000 miles! Imagine the earth and the other planets constituting the solar system removed to Arcturus and set revolving around it in orbits of the same forms and sizes as those in which they circle about the sun. Poor Mercury! For that little planet it would indeed be a jump from the frying pan into the fire, because, as it rushed to perihelion, Mercury would plunge more than 2,500,000 miles beneath the surface of the giant star. Venus and the earth would melt like snowflakes at the mouth of a furnace. Even far-away Neptune, the remotest member of the system, would swelter in torrid heat.

But stop! Look at the sky. Observe how small and motionless the disks of the stars have become. Back to the telescopes at once, for this is a token that the atmosphere is steady, and that "good seeing" may be expected. It is fortunate, for we have some delicate work before us. The very first double star we try in Sigma 1772, requires the use of the four-inch, and the five-inch shows it more satisfactorily. The magnitudes are sixth and ninth, distance 5", p. 140? On the other side of Arcturus we find zeta, a star that we should have had no great difficulty in separating thirty years ago, but which has now closed up beyond the reach even of our five-inch. The magnitudes are both fourth, and the distance less than a quarter of a second; position angle changing. It is apparently a binary, and if so will some time widen again, but its period is unknown. The star 279, also known as Sigma 1910, near the southeastern edge of the constellation, is a pretty double, each component being of the seventh magnitude, distance 4", p. 212? Just above zeta we come upon pi, an easy double for the three-inch, magnitudes four and six, distance 6" p. 99? Next is xi, a yellow and purple pair, whose magnitudes are respectively five and seven, distance less than 3", p. 200? This is undoubtedly a binary with a period of revolution of about a hundred and thirty years. Its distance decreased about 1" between 1881 and 1891. It was still decreasing in 1899, when it had become 2.5". The orbital swing is also very apparent in the change of the position angle.

The telescopic gem and one of "the flowers of the sky," is epsilon, also known as Mirac. When well seen, as we shall see it to-night, epsilon is superb. The magnitudes of its two component stars are two and a half (according to Hall, three) and six. The distance is about 2.8", p. 326? The contrast of colors--bright orange yellow, set against brilliant emerald green--is magnificent. There are very few doubles that can be compared with it in this respect. The three-inch will separate it, but the five-inch enables us best to enjoy its beauty. It appears to be a binary, but the motion is very slow, and nothing certain is yet known of its period.

In delta we have a very wide and easy double; magnitudes three and a half and eight and a half, distance 110", p. 75? The smaller star has a lilac hue. We can not hope with any of our instruments to see all of the three stars contained in , but two of them are easily seen; magnitudes four and seven, distance 108", p. 172? The smaller star is again double; magnitudes seven and eight, distance 0.77", p. 88? It is clearly a binary, with a long period. A six-inch telescope that could separate this star at present would be indeed a treasure. Sigma 1926 is another object rather beyond our powers, on account of the contrast of magnitudes. These are six and eight and a half; distance 1.3", p. 256?

Other doubles are: 44 (Sigma 1909), magnitudes five and six, distance 4.8", p. 240? 39 (Sigma 1890), magnitudes both nearly six, distance 3.6", p. 45? Smaller star light red; iota, magnitudes four and a half and seven and a half, distance 38", p. 33? kappa, magnitudes five and a half and eight, distance 12.7", p. 238? Some observers see a greenish tinge in the light of the larger star, the smaller one being blue.

There are one or two interesting things to be seen in that part of Canes Venatici which is represented on map No. 11. The first of these is the star cluster 3936. This will reward a good look with the five-inch. With large telescopes as many as one thousand stars have been discerned packed within its globular outlines.

The star 25 (Sigma 1768) is a close binary with a period estimated at one hundred and twenty-five years. The magnitudes are six and seven or eight, distance about 1", p. 137? We may try for this with the five-inch, and if we do not succeed in separating the stars we may hope to do so some time, for the distance between them is increasing.

Although the nebula 3572 is a very wonderful object, we shall leave it for another evening.

Eastward shines the circlet of Corona Borealis, whose form is so strikingly marked out by the stars that the most careless eye perceives it at once. Although a very small constellation, it abounds with interesting objects. We begin our attack with the five-inch on Sigma 1932, but not too confident that we shall come off victors, for this binary has been slowly closing for many years. The magnitudes are six and a half and seven, distance 0.84", p. 150? Not far distant is another binary, at present beyond our powers, eta. Here the magnitudes are both six, distance 0.65", p. 3? Hall assigns a period of forty years to this star.

The assemblage of close binaries in this neighborhood is very curious. Only a few degrees away we find one that is still more remarkable, the star gamma. What has previously been said about 42 Com Berenicis applies in a measure to this star also. It, too, has a comparatively small orbit, and its components are never seen widely separated. In 1826 their distance was 0.7"; in 1880 they could not be split; in 1891 the distance had increased to 0.36", and in 1894 it had become 0.53", p. 123? But in 1899 Lewis made the distance only 0.43". The period has been estimated at one hundred years.

While the group of double stars in the southern part of Corona Borealis consists, as we have seen, of remarkably close binaries, another group in the northern part of the same constellation comprises stars that are easily separated. Let us first try zeta. The powers of the three-inch are amply sufficient in this case. The magnitudes are four and five, distance 6.3", p. 300? Colors, white or bluish-white and blue or green.

Next take sigma, whose magnitudes are five and six, distance 4", p. 206? With the five-inch we may look for a second companion of the tenth magnitude, distance 54", p. 88? It is thought highly probable that sigma is a binary, but its period has simply been guessed at.

Finally, we come to nu, which consists of two very widely separated stars, nu^1 and nu^2, each of which has a faint companion. With the five-inch we may be able to see the companion of nu^2, the more southerly of the pair. The magnitude of the companion is variously given as tenth and twelfth, distance 137", p. 18?

With the aid of the map we find the position of the new star of 1866, which is famous as the first so-called temporary star to which spectroscopic analysis was applied. When first noticed, on May 12, 1866, this star was of the second magnitude, fully equaling in brilliancy alpha, the brightest star of the constellation; but in about two weeks it fell to the ninth magnitude. Huggins and Miller eagerly studied the star with the spectroscope, and their results were received with deepest interest. They concluded that the light of the new star had two different sources, each giving a spectrum peculiar to itself. One of the spectra had dark lines and the other bright lines. It will be remembered that a similar peculiarity was exhibited by the new star in Auriga in 1893. But the star in Corona did not disappear. It diminished to magnitude nine and a half or ten, and stopped there; and it is still visible. In fact, subsequent examination proved that it had been catalogued at Bonn as a star of magnitude nine and a half in 1855. Consequently this "blaze star" of 1866 will bear watching in its decrepitude. Nobody knows but that it may blaze again. Perhaps it is a sun-like body; perhaps it bears little resemblance to a sun as we understand such a thing. But whatever it may be, it has proved itself capable of doing very extraordinary things.

We have no reason to suspect the sun of any latent eccentricities, like those that have been displayed by "temporary" stars; yet, acting on the principle which led the old emperor-astrologer Rudolph II to torment his mind with self-made horoscopes of evil import, let us unscientifically imagine that the sun could suddenly burst out with several hundred times its ordinary amount of heat and light, thereby putting us into a proper condition for spectroscopic examination by curious astronomers in distant worlds.

But no, after all, it is far pleasanter to keep within the strict boundaries of science, and not imagine anything of the kind.

CHAPTER V

IN SUMMER STAR-LANDS

"I heard the trailing garments of the night Sweep through her marble halls, I saw her sable skirts all fringed with light From the celestial walls."--H. W. LONGFELLOW.

In the soft air of a summer night, when fireflies are flashing their lanterns over the fields, the stars do not sparkle and blaze like those that pierce the frosty skies of winter. The light of Sirius, Aldebaran, Rigel, and other midwinter brilliants possesses a certain gemlike hardness and cutting quality, but Antares and Vega, the great summer stars, and Arcturus, when he hangs westering in a July night, exhibit a milder radiance, harmonizing with the character of the season. This difference is, of course, atmospheric in origin, although it may be partly subjective, depending upon the mental influences of the mutations of Nature.

The constellation Scorpio is nearly as striking in outline as Orion, and its brightest star, the red Antares (alpha in map No. 12), carries concealed in its rays a green jewel which, to the eye of the enthusiast in telescopic recreation, appears more beautiful and inviting each time that he penetrates to its hiding place.

We shall begin our night's work with this object, and the four-inch glass will serve our purpose, although the untrained observer would be more certain of success with the five-inch. A friend of mine has seen the companion of Antares with a three-inch, but I have never tried the star with so small an aperture. When the air is steady and the companion can be well viewed, there is no finer sight among the double stars. The contrast of colors is beautifully distinct--fire-red and bright green. The little green star has been seen emerging from behind the moon, ahead of its ruddy companion. The magnitudes are one and seven and a half or eight, distance 3", p. 270? Antares is probably a binary, although its binary character has not yet been established.

A slight turn of the telescope tube brings us to the star sigma, a wide double, the smaller component of which is blue or plum-colored; magnitudes four and nine, distance 20", p. 272? From sigma we pass to beta, a very beautiful object, of which the three-inch gives us a splendid view. Its two components are of magnitudes two and six, distance 13", p. 30? colors, white and bluish. It is interesting to know that the larger star is itself double, although none of the telescopes we are using can split it. Burnham discovered that it has a tenth-magnitude companion; distance less than 1", p. 87?

And now for a triple, which will probably require the use of our largest glass. Up near the end of the northern prolongation of the constellation we perceive the star xi. The three-inch shows us that it is double; the five-inch divides the larger star again. The magnitudes are respectively five, five and a half, and seven and a half, distances 0.94", p. 215? and 7", p. 70?

A still more remarkable star, although one of its components is beyond our reach, is nu. With the slightest magnifying this object splits up into two stars, of magnitudes four and seven, situated rather more than 40" apart. A high power divides the seventh-magnitude companion into two, each of magnitude six and a half, distance 1.8", p. 42? But (and this was another of Burnham's discoveries) the fourth-magnitude star itself is double, distance 0.8", p. about 0? The companion in this case is of magnitude five and a half.

Next we shall need a rather low-power eyepiece and our largest aperture in order to examine a star cluster, No. 4173, which was especially admired by Sir William Herschel, who discovered that it was not, as Messier had supposed, a circular nebula. Herschel regarded it as the richest mass of stars in the firmament, but with a small telescope it appears merely as a filmy speck that has sometimes been mistaken for a comet. In 1860 a new star, between the sixth and seventh magnitude in brilliance, suddenly appeared directly in or upon the cluster, and the feeble radiance of the latter was almost extinguished by the superior light of the stranger. The latter disappeared in less than a month, and has not been seen again, although it is suspected to be a variable, and, as such, has been designated with the letter T. Two other known variables, both very faint, exist in the immediate neighborhood. According to the opinion that was formerly looked upon with favor, the variable T, if it is a variable, simply lies in the line of sight between the earth and the star cluster, and has no actual connection with the latter. But this opinion may not, after all, be correct, for Mr. Bailey's observations show that variable stars sometimes exist in large numbers in clusters, although the variables thus observed are of short period. The cluster 4183, just west of Antares, is also worth a glance with the five-inch glass. It is dense, but its stars are very small, so that to enjoy its beauty we should have to employ a large telescope. Yet there is a certain attraction in these far-away glimpses of starry swarms, for they give us some perception of the awful profundity of space. When the mind is rightly attuned for these revelations of the telescope, there are no words that can express its impressions of the overwhelming perspective of the universe.

The southern part of the constellation Ophiuchus is almost inextricably mingled with Scorpio. We shall, therefore, look next at its attractions, beginning with the remarkable array of star clusters 4264, 4268, 4269, and 4270. All of these are small, 2' or 3' in diameter, and globular in shape. No. 4264 is the largest, and we can see some of the stars composing it. But these clusters, like those just described in Scorpio, are more interesting for what they signify than for what they show; and the interest is not diminished by the fact that their meaning is more or less of a mystery. Whether they are composed of pygmy suns or of great solar globes like that one which makes daylight for the earth, their association in spherical groups is equally suggestive.

There are two other star clusters in Ophiuchus, and within the limits of map No. 12, both of which are more extensive than those we have just been looking at. No. 4211 is 5' or 6' in diameter, also globular, brighter at the center, and surrounded by several comparatively conspicuous stars. No. 4346 is still larger, about half as broad as the moon, and many of its scattered stars are of not less than the ninth magnitude. With a low magnifying power the field of view surrounding the cluster appears powdered with stars.

There are only two noteworthy doubles in that part of Ophiuchus with which we are at present concerned: 36, whose magnitudes are five and seven, distance 4.3", p. 195? colors yellow and red; and 39, magnitudes six and seven and a half, distance 12", p. 356? colors yellow or orange and blue. The first named is a binary whose period has not been definitely ascertained.

The variable R has a period a little less than three hundred and three days. At its brightest it is of magnitude seven or eight, and at minimum it diminishes to about the twelfth magnitude.

The spot where the new star of 1604 appeared is indicated on the map. This was, with the exception of Tycho's star in 1572, the brightest temporary star of which we possess a trustworthy account. It is frequently referred to as Kepler's star, because Kepler watched it with considerable attention, but unfortunately he was not as good an observer as Tycho was. The star was first seen on October 10, 1604, and was then brighter than Jupiter. It did not, however, equal Venus. It gradually faded and in March, 1606, disappeared. About twelve degrees northwest of the place of the star of 1604, and in that part of the constellation Serpens which is included in map No. 12, we find the location of another temporary star, that of 1848. It was first noticed by Mr. Hind on April 28th of that year, when its magnitude was not much above the seventh, and its color was red. It brightened rapidly, until on May 2d it was of magnitude three and a half. Then it began to fade, but very slowly, and it has never entirely disappeared. It is now of the twelfth or thirteenth magnitude.

In passing we may glance with a low power at nu Serpentis, a wide double, magnitudes four and nine, distance 50", p. 31? colors contrasted but uncertain.

Sagittarius and its neighbor, the small but rich constellation Scutum Sobieskii, attract us next. We shall first deal with the western portions of these constellations which are represented on Map No. 12. The star in Sagittarius is a wide triple, magnitudes three and a half, nine and a half, and ten, distances 40", p. 315? and 45", p. 114? But the chief glory of Sagittarius (and the same statement applies to Scutum Sobieskii) lies in its assemblage of star clusters. One of these, No. 4361, also known as M 8, is plainly visible to the naked eye as a bright spot in the Milky Way. We turn our five-inch telescope, armed with a low magnifying power, upon this subject and enjoy a rare spectacle. As we allow it to drift through the field we see a group of three comparatively brilliant stars advancing at the front of a wonderful train of mingled star clusters and nebulous clouds. A little northwest of it appears the celebrated trifid nebula, No. 4355 on the map. There is some evidence that changes have occurred in this nebula since its discovery in the last century. Barnard has made a beautiful photograph showing M 8 and the trifid nebula on the same plate, and he remarks that the former is a far more remarkable object than its more famous neighbor. Near the eastern border of the principal nebulous cloud there is a small and very black hole with a star poised on its eastern edge. This hole and the star are clearly shown in the photograph.

Cluster No. 4397 (M 24) is usually described as resembling, to the naked eye, a protuberance on the edge of the Milky Way. It is nearly three times as broad as the moon, and is very rich in minute stars, which are at just such a degree of visibility that crowds of them continually appear and disappear while the eye wanders over the field, just as faces are seen and lost in a vast assemblage of people. This kind of luminous agitation is not peculiar to M 24, although that cluster exhibits it better than most others do on account of both the multitude and the minuteness of its stars.

A slight sweep eastward brings us to yet another meeting place of stars, the cluster M 25, situated between the variables U and V. This is brilliant and easily resolved into its components, which include a number of double stars.

The two neighboring variables just referred to are interesting. U has a period of about six days and three quarters, and its range of magnitude runs from the seventh down to below the eighth. V is a somewhat mysterious star. Chandler removed it from his catalogue of variables because no change had been observed in its light by either himself, Sawyer, or Yendell. Quirling, the discoverer of its variability, gave the range as between magnitudes 7.6 and 8.8. It must, therefore, be exceedingly erratic in its changes, resembling rather the temporary stars than the true variables.

In that part of Scutum Sobieskii contained in map No. 12 we find an interesting double, Sigma 2325, whose magnitudes are six and nine, distance 12.3", p. 260? colors white and orange. Sigma 2306 is a triple, magnitudes seven, eight, and nine, distances 12", p. 220? and 0.8", p. 68? The third star is, however, beyond our reach. The colors of the two larger are respectively yellow and violet.

The star cluster 4400 is about one quarter as broad as the moon, and easily seen with our smallest aperture.

Passing near to the region covered by map No. 13, we find the remaining portions of the constellations Sagittarius and Scutum Sobieskii. It will be advisable to finish with the latter first. Glance at the clusters 4426 and 4437. Neither is large, but both are rich in stars. The nebula 4441 is a fine object of its kind. It brightens toward the center, and Herschel thought he had resolved it into stars. The variable R is remarkable for its eccentricities. Sometimes it attains nearly the fourth magnitude, although usually at maximum it is below the fifth, while at minimum it is occasionally of the sixth and at other times of the seventh or eighth magnitude. Its period is irregular.

Turning back to Sagittarius, we resume our search for interesting objects there, and the first that we discover is another star cluster, for the stars are wonderfully gregarious in this quarter of the heavens. The number our cluster bears on the map is 4424, corresponding with M 22 in Messier's catalogue. It is very bright, containing many stars of the tenth and eleventh magnitudes, as well as a swarm of smaller ones. Sir John Herschel regarded the larger stars in this cluster as possessing a reddish tint. Possibly there was some peculiarity in his eye that gave him this impression, for he has described a cluster in the constellation Toucan in the southern hemisphere as containing a globular mass of rose-colored stars inclosed in a spherical shell of white stars. Later observers have confirmed his description of the shape and richness of this cluster in Toucan, but have been unable to perceive the red hue of the interior stars.

The eastern expanse of Sagittarius is a poor region compared with the western end of the constellation, where the wide stream of the Milky Way like a great river enriches its surroundings. The variables T and R are of little interest to us, for they never become bright enough to be seen without the aid of a telescope. In 54 we find, however, an interesting double, which with larger telescopes than any of ours appears as a triple. The two stars that we see are of magnitudes six and seven and a half, distance 45", p. 42? colors yellow and blue. The third star, perhaps of thirteenth magnitude, is distant 36", p. 245?

Retaining map No. 13 as our guide, we examine the western part of the constellation Capricornus. Its leader alpha is a naked-eye double, the two stars being a little more than 6' apart. Their magnitudes are three and four, and both have a yellowish hue. The western star is alpha^1, and is the fainter of the two. The other is designated as alpha^2. Both are double. The components of alpha^1 are of magnitudes four and eight and a half, distance 44", p. 220? With the Washington twenty-six-inch telescope a third star of magnitude fourteen has been found at a distance of 40", p. 182? In alpha^2 the magnitudes of the components are three and ten and a half, distance 7.4", p. 150? The smaller star has a companion of the twelfth or thirteenth magnitude, distance 1.2", p. 240? This, of course, is hopelessly beyond our reach. Yet another star of magnitude nine, distance 154", p. 156, we may see easily.

Dropping down to beta, we find it to be a most beautiful and easy double, possessing finely contrasted colors, gold and blue. The larger star is of magnitude three, and the smaller, the blue one, of magnitude six, distance 205", p. 267? Between them there is a very faint star which larger telescopes than ours divide into two, each of magnitude eleven and a half; separated 3", p. 325?

Still farther south and nearly in a line drawn from alpha through beta we find a remarkable group of double stars, sigma, pi, rho, and omicron. The last three form a beautiful little triangle. We begin with sigma, the faintest of the four. The magnitudes of its components are six and nine, distance 54", p. 177? In pi the magnitudes are five and nine, distance 3.4", p. 145? in rho, magnitudes five and eight, distance 3.8", p. 177?(a third star of magnitude seven and a half is seen at a distance of 4', p. 150?; in omicron, magnitudes six and seven, distance 22", p. 240?

The star cluster 4608 is small, yet, on a moonless night, worth a glance with the five-inch.

We now pass northward to the region covered by map No. 14, including the remainder of Ophiuchus and Serpens. Beginning with the head of Serpens, in the upper right-hand corner of the map, we find that beta, of magnitude three and a half, has a ninth-magnitude companion, distance 30", p. 265? The larger star is light blue and the smaller one yellowish. The little star nu is double, magnitudes five and nine, distance 50", p. 31? colors contrasted but uncertain. In delta we find a closer double, magnitudes three and four, distance 3.5", p. 190? It is a beautiful object for the three-inch. The leader of the constellation, alpha, of magnitude two and a half, has a faint companion of only the twelfth magnitude, distance 60", p. 350? The small star is bluish. The variable R has a period about a week short of one year, and at maximum exceeds the sixth magnitude, although sinking at minimum to less than the eleventh. Its color is ruddy.

Passing eastward, we turn again into Ophiuchus, and find immediately the very interesting double, lambda, whose components are of magnitudes four and six, distance 1", p. 55? This is a long-period binary, and notwithstanding the closeness of its stars, our four-inch should separate them when the seeing is fine. We shall do better, however, to try with the five-inch. Sigma 2166 consists of two stars of magnitudes six and seven and a half, distance 27", p. 280? Sigma 2173 is a double of quite a different order. The magnitudes of its components are both six, the distance in 1899 0.98", p. 331? It is evidently a binary in rapid motion, as the distance changed from about a quarter of a second in 1881 to more than a second in 1894. The star tau is a fine triple, magnitudes five, six, and nine, distances 1.8", p. 254? and 100", p. 127? The close pair is a binary system with a long period of revolution, estimated at about two hundred years. We discover another group of remarkable doubles in 67, 70, and 73. In the first-named star the magnitudes are four and eight, distance 55", p. 144? colors finely contrasted, pale yellow and red.

Much more interesting, however, is 70, a binary whose components have completed a revolution since their discovery by Sir William Herschel, the period being ninety-five years. The magnitudes are four and six, or, according to Hall, five and six, distance in 1894 2.3"; in 1900, 1.45", according to Maw. Hall says the apparent distance when the stars are closest is about 1.7", and when they are widest 6.7". This star is one of those whose parallax has been calculated with a reasonable degree of accuracy. Its distance from us is about 1,260,000 times the distance of the sun, the average distance apart of the two stars is about 2,800,000,000 miles (equal to the distance of Neptune from the sun), and their combined mass is three times that of the sun. Hall has seen in the system of 70 Ophiuchi three stars of the thirteenth magnitude or less, at distances of about 60", 90", and 165" respectively.

The star 73 is also a close double, and beyond our reach. Its magnitudes are six and seven, distance 0.7", p. 245? It is, no doubt, a binary.

Three star clusters in Ophiuchus remain to be examined. The first of these, No. 4256, is partially resolved into stars by the five-inch. No. 4315 is globular, and has a striking environment of bystanding stars. It is about one quarter as broad as the full moon, and our largest aperture reveals the faint coruscation of its crowded components. No. 4410 is a coarser and more scattered star swarm--a fine sight!

Farther toward the east we encounter a part of Serpens again, which contains just one object worth glancing at, the double theta, whose stars are of magnitudes four and four and a half, distance 21", p. 104? Color, both yellow, the smaller star having the deeper hue.

Let us next, with the guidance of map No. 15, enter the rich star fields of Hercules, and of the head and first coils of Draco. According to Argelander, Hercules contains more stars visible to the naked eye than any other constellation, and he makes the number of them one hundred and fifty-five, nearly two thirds of which are only of the sixth magnitude. But Heis, who saw more naked-eye stars than Argelander, makes Ursa Major precisely equal to Hercules in the number of stars, his enumeration showing two hundred and twenty-seven in each constellation, while, according to him, Draco follows very closely after, with two hundred and twenty stars. Yet, on account of the minuteness of the majority of their stars, neither of these constellations makes by any means as brilliant a display as does Orion, to which Argelander assigns only one hundred and fifteen naked-eye stars, and Heis one hundred and thirty-six.

We begin in Hercules with the star kappa, a pretty little double of magnitudes five and a half and seven, distance 31", p. 10? colors yellow and red. Not far away we find, in gamma, a larger star with a fainter companion, the magnitudes in this case being three and a half and nine, distance 38", p. 242? colors white and faint blue or lilac. One of the most beautiful of double stars is alpha Herculis. The magnitudes are three and six, distance 4.7", p. 118? colors orange and green, very distinct. Variability has been ascribed to each of the stars in turn. It is not known that they constitute a binary system, because no certain evidence of motion has been obtained. Another very beautiful and easily separated double is delta, magnitudes three and eight, distance 19", p. 175? colors pale green and purple.

Sweeping northwestward to zeta, we encounter a celebrated binary, to separate which at present requires the higher powers of a six-inch glass. The magnitudes are three and six and a half, distance in 1899, 0.6", p. 264? in 1900, 0.8", p. 239? The period of revolution is thirty-five years, and two complete revolutions have been observed. The apparent distance changes from 0.6" to 1.6". They were at their extreme distance in 1884.

Two pleasing little doubles are Sigma 2101, magnitudes six and nine, distance 4", p. 57? and Sigma 2104, magnitudes six and eight, distance 6", p. 20? At the northern end of the constellation is 42, a double that requires the light-grasping power of our largest glass. Its magnitudes are six and twelve, distance 20", p. 94? In rho we discover another distinctly colored double, both stars being greenish or bluish, with a difference of tone. The magnitudes are four and five and a half, distance 3.7", p. 309? But the double 95 is yet more remarkable for the colors of its stars. Their magnitudes are five and five and a half, distance 6", p. 262? colors, according to Webb, "light apple-green and cherry-red." But other observers have noted different hues, one calling them both golden yellow. I think Webb's description is more nearly correct. Sigma 2215 is a very close double, requiring larger telescopes than those we are working with. Its magnitudes are six and a half and eight, distance 0.7", p. 300? It is probably a binary. Sigma 2289 is also close, but our five-inch will separate it: magnitudes six and seven, distance 1.2", p. 230?

Turning to , we have to deal with a triple, one of whose stars is at present beyond the reach of our instruments. The magnitudes of the two that we see are four and ten, distance 31", p. 243? The tenth-magnitude star is a binary of short period (probably less than fifty years), the distance of whose components was 2" in 1859, 1" in 1880, 0.34" in 1889, and 0.54" in 1891, when the position angle was 25? and rapidly increasing. The distance is still much less than 1".

For a glance at a planetary nebula we may turn with the five-inch to No. 4234. It is very small and faint, only 8" in diameter, and equal in brightness to an eighth-magnitude star. Only close gazing shows that it is not sharply defined like a star, and that it possesses a bluish tint. Its spectrum is gaseous.

The chief attraction of Hercules we have left for the last, the famous star cluster between eta and zeta, No. 4230, more commonly known as M 13. On a still evening in the early summer, when the moon is absent and the quiet that the earth enjoys seems an influence descending from the brooding stars, the spectacle of this sun cluster in Hercules, viewed with a telescope of not less than five-inches aperture, captivates the mind of the most uncontemplative observer. With the Lick telescope I have watched it resolve into separate stars to its very center--a scene of marvelous beauty and impressiveness. But smaller instruments reveal only the in-running star streams and the sprinkling of stellar points over the main aggregation, which cause it to sparkle like a cloud of diamond dust transfused with sunbeams. The appearance of flocking together that those uncountable thousands of stars present calls up at once a picture of our lone sun separated from its nearest stellar neighbor by a distance probably a hundred times as great as the entire diameter of the spherical space within which that multitude is congregated. It is true that unless we assume what would seem an unreasonable remoteness for the Hercules cluster, its component stars must be much smaller bodies than the sun; yet even that fact does not diminish the wonder of their swarming. Here the imagination must bear science on its wings, else science can make no progress whatever. It is an easy step from Hercules to Draco. In the conspicuous diamond-shaped figure that serves as a guide-board to the head of the latter, the southernmost star belongs not to Draco but to Hercules. The brightest star in this figure is gamma, of magnitude two and a half, with an eleventh-magnitude companion, distant 125", p. 116? Two stars of magnitude five compose nu, their distance apart being 62", p. 312? A more interesting double is , magnitudes five and five, distance 2.4", p. 158? Both stars are white, and they present a pretty appearance when the air is steady. They form a binary system of unknown period. Sigma 2078 (also called 17 Draconis) is a triple, magnitudes six, six and a half, and six, distances 3.8", p. 116? and 90", p. 195? Sigma 1984 is an easy double, magnitudes six and a half and eight and a half, distance 6.4", p. 276? The star eta is a very difficult double for even our largest aperture, on account of the faintness of one of its components. The magnitudes are two and a half and ten, distance 4.7", p. 140? Its near neighbor, Sigma 2054, may be a binary. Its magnitudes are six and seven, distance 1", p. 0? In Sigma 2323 we have another triple, magnitudes five, eight and a half, and seven, distances 3.6", p. 360? and 90", p. 22? colors white, blue, and reddish. A fine double is epsilon, magnitudes five and eight, distance 3", p. 5?

The nebula No. 4373 is of a planetary character, and interesting as occupying the pole of the ecliptic. A few years ago Dr. Holden, with the Lick telescope, discovered that it is unique in its form. It consists of a double spiral, drawn out nearly in the line of sight, like the thread of a screw whose axis lies approximately endwise with respect to the observer. There is a central star, and another fainter star is involved in the outer spiral. The form of this object suggests strange ideas as to its origin. But the details mentioned are far beyond the reach of our instruments. We shall only see it as a hazy speck. No. 4415 is another nebula worth glancing at. It is Tuttle's so-called variable nebula.

There are three constellations represented on map No. 16 to which we shall pay brief visits. First Aquila demands attention. Its doubles may be summarized as follows: 11, magnitudes five and nine, distance 17.4", p. 252? pi, magnitudes six and seven, distance 1.6", p. 122? 23, magnitudes six and ten, distance 3.4", p. 12?-requires the five-inch and good seeing; 57, magnitudes five and six, distance 36", p. 170? Sigma 2654, magnitudes six and eight, distance 12", p. 234? Sigma 2644, magnitudes six and seven, distance 3.6", p. 208?

The star eta is an interesting variable between magnitudes three and a half and 4.7; period, seven days, four hours, fourteen minutes. The small red variable R changes from magnitude six to magnitude seven and a half and back again in a period of three hundred and fifty-one days.

Star cluster No. 4440 is a striking object, its stars ranging from the ninth down to the twelfth magnitude.

Just north of Aquila is the little constellation Sagitta, containing several interesting doubles and many fine star fields, which may be discovered by sweeping over it with a low-power eyepiece. The star zeta is double, magnitudes five and nine, distance 8.6", p. 312? The larger star is itself double, but far too close to be split, except with very large telescopes. In theta we find three components of magnitudes seven, nine, and eight respectively, distances 11.4", p. 327? and 70", p. 227? A wide double is epsilon, magnitudes six and eight, distance 92", p. 81? Nebula No. 4572 is planetary.

Turning to Delphinus, we find a very beautiful double in gamma, magnitudes four and five, distance 11", p. 273? colors golden and emerald. The leader alpha, which is not as bright as its neighbor beta, and which is believed to be irregularly variable, is of magnitude four, and has a companion of nine and a half magnitude at the distance 35", p. 278? At a similar distance, 35", p. 335? beta has an eleventh-magnitude companion, and the main star is also double, but excessively close, and much beyond our reach. It is believed to be a swiftly moving binary, whose stars are never separated widely enough to be distinguished with common telescopes.

CHAPTER VI

FROM LYRA TO ERIDANUS

"This Orpheus struck when with his wondrous song He charmed the woods and drew the rocks along."--MANILIUS.

We resume our celestial explorations with the little constellation Lyra, whose chief star, Vega (alpha), has a very good claim to be regarded as the most beautiful in the sky. The position of this remarkable star is indicated in map No. 17. Every eye not insensitive to delicate shades of color perceives at once that Vega is not white, but blue-white. When the telescope is turned upon the star the color brightens splendidly. Indeed, some glasses decidedly exaggerate the blueness of Vega, but the effect is so beautiful that one can easily forgive the optical imperfection which produces it. With our four-inch we look for the well-known companion of Vega, a tenth-magnitude star, also of a blue color deeper than the hue of its great neighbor. The distance is 50", p. 158? Under the most favorable circumstances it might be glimpsed with the three-inch, but, upon the whole, I should regard it as too severe a test for so small an aperture.

Vega is one of those stars which evidently are not only enormously larger than the sun (one estimate makes the ratio in this case nine hundred to one), but whose physical condition, as far as the spectroscope reveals it, is very different from that of our ruling orb. Like Sirius, Vega displays the lines of hydrogen most conspicuously, and it is probably a much hotter as well as a much more voluminous body than the sun.

Close by, toward the east, two fourth-magnitude stars form a little triangle with Vega. Both are interesting objects for the telescope, and the northern one, epsilon, has few rivals in this respect. Let us first look at it with an opera glass. The slight magnifying power of such an instrument divides the star into two twinkling points. They are about two and a quarter minutes of arc apart, and exceptionally sharp-sighted persons are able to see them divided with the naked eye. Now take the three-inch telescope and look at them, with a moderate power. Each of the two stars revealed by the opera glass appears double, and a fifth star of the ninth magnitude is seen on one side of an imaginary line joining the two pairs. The northern-most pair is named epsilon1, the magnitudes being fifth and sixth, distance 3", p. 15? The other pair is epsilon2, magnitudes fifth and sixth, distance 2.3", p. 133? Each pair is apparently a binary; but the period of revolution is unknown. Some have guessed a thousand years for one pair, and two thousand for the other. Another guess gives epsilon1 a period of one thousand years, and epsilon2 a period of eight hundred years. Hall, in his double-star observations, simply says of each, "A slow motion."

Purely by guesswork a period has also been assigned to the two pairs in a supposed revolution around their common center, the time named being about a million years. It is not known, however, that such a motion exists. Manifestly it could not be ascertained within the brief period during which scientific observations of these stars have been made. The importance of the element of time in the study of stellar motions is frequently overlooked, though not, of course, by those who are engaged in such work. The sun, for instance, and many of the stars are known to be moving in what appear to be straight lines in space, but observations extending over thousands of years would probably show that these motions are in curved paths, and perhaps in closed orbits.

If now in turn we take our four-inch glass, we shall see something else in this strange family group of epsilon Lyr? Between epsilon1 and epsilon2, and placed one on each side of the joining line, appear two exceedingly faint specks of light, which Sir John Herschel made famous under the name of the debilissima. They are of the twelfth or thirteenth magnitude, and possibly variable to a slight degree. If you can not see them at first, turn your eye toward one side of the field of view, and thus, by bringing their images upon a more sensitive part of the retina, you may glimpse them. The sight is not much, yet it will repay you, as every glance into the depths of the universe does.

The other fourth-magnitude star near Vega is zeta, a wide double, magnitudes fourth and sixth, distance 44", p. 150? Below we find beta, another very interesting star, since it is both a multiple and an eccentric variable. It has four companions, three of which we can easily see with our three-inch; the fourth calls for the five-inch; the magnitudes are respectively four, seven or under, eight, eight and a half, and eleven; distances 45", p. 150? 65", p. 320? 85", p. 20? and 46", p. 248? The primary, beta, varies from about magnitude three and a half to magnitude four and a half, the period being twelve days, twenty-one hours, forty-six minutes, and fifty-eight seconds. Two unequal maxima and minima occur within this period. In the spectrum of this star some of the hydrogen lines and the D3 line (the latter representing helium, a constituent of the sun and of some of the stars, which, until its recent discovery in a few rare minerals was not known to exist on the earth) are bright, but they vary in visibility. Moreover, dark lines due to hydrogen also appear in its spectrum simultaneously with the bright lines of that element. Then, too, the bright lines are sometimes seen double. Professor Pickering's explanation is that beta Lyr?probably consists of two stars, which, like the two composing beta Aurig? are too close to be separated with any telescope now existing, and that the body which gives the bright lines is revolving in a circle in a period of about twelve days and twenty-two hours around the body which gives the dark lines. He has also suggested that the appearances could be accounted for by supposing a body like our sun to be rotating in twelve days and twenty-two hours, and having attached to it an enormous protuberance extending over more than one hundred and eighty degrees of longitude, so that when one end of it was approaching us with the rotation of the star the other end would be receding, and a splitting of the spectral lines at certain periods would be the consequence. "The variation in light," he adds, "may be caused by the visibility of a larger or smaller portion of this protuberance."

Unfortunate star, doomed to carry its parasitical burden of hydrogen and helium, like Sindbad in the clasp of the Old Man of the Sea! Surely, the human imagination is never so wonderful as when it bears an astronomer on its wings. Yet it must be admitted that the facts in this case are well calculated to summon the genius of hypothesis. And the puzzle is hardly simplified by B 閘 opolsky's observation that the body in beta Lyr?giving dark hydrogen lines shows those lines also split at certain times. It has been calculated, from a study of the phenomena noted above, that the bright-line star in beta Lyr?is situated at a distance of about fifteen million miles from the center of gravity of the curiously complicated system of which it forms a part.

We have not yet exhausted the wonders of Lyra. On a line from beta to gamma, and about one third of the distance from the former to the latter, is the celebrated Ring Nebula, indicated on the map by the number 4447. We need all the light we can get to see this object well, and so, although the three-inch will show it, we shall use the five-inch. Beginning with a power of one hundred diameters, which exhibits it as a minute elliptical ring, rather misty, very soft and delicate, and yet distinct, we increase the magnification first to two hundred and finally to three hundred, in order to distinguish a little better some of the details of its shape. Upon the whole, however, we find that the lowest power that clearly brings out the ring gives the most satisfactory view. The circumference of the ring is greater than that of the planet Jupiter. Its ellipticity is conspicuous, the length of the longer axis being 78" and that of the shorter 60". Closely following the nebula as it moves through the field of view, our five-inch telescope reveals a faint star of the eleventh or twelfth magnitude, which is suspected of variability. The largest instruments, like the Washington and the Lick glasses, have shown perhaps a dozen other stars apparently connected with the nebula. A beautiful sparkling effect which the nebula presents was once thought to be an indication that it was really composed of a circle of stars, but the spectroscope shows that its constitution is gaseous. Just in the middle of the open ring is a feeble star, a mere spark in the most powerful telescope. But when the Ring Nebula is photographed--and this is seen beautifully in the photographs made with the Crossley reflector on Mount Hamilton by the late Prof. J. E. Keeler--this excessively faint star imprints its image boldly as a large bright blur, encircled by the nebulous ring, which itself appears to consist of a series of intertwisted spirals.

Not far away we find a difficult double star, 17, whose components are of magnitudes six and ten or eleven, distance 3.7", p. 325?

From Lyra we pass to Cygnus, which, lying in one of the richest parts of the Milky Way, is a very interesting constellation for the possessor of a telescope. Its general outlines are plainly marked for the naked eye by the figure of a cross more than twenty degrees in length lying along the axis of the Milky Way. The foot of the cross is indicated by the star beta, also known as Albireo, one of the most charming of all the double stars. The three-inch amply suffices to reveal the beauty of this object, whose components present as sharp a contrast of light yellow and deep blue as it would be possible to produce artificially with the purest pigments. The magnitudes are three and seven, distance 34.6", p. 55? No motion has been detected indicating that these stars are connected in orbital revolution, yet no one can look at them without feeling that they are intimately related to one another. It is a sight to which one returns again and again, always with undiminished pleasure. The most inexperienced observer admires its beauty, and after an hour spent with doubtful results in trying to interest a tyro in double stars it is always with a sense of assured success that one turns the telescope to beta Cygni.

Following up the beam of the imaginary cross along the current of the Milky Way, every square degree of which is here worth long gazing into, we come to a pair of stars which contend for the name-letter chi. On our map the letter is attached to the southernmost of the two, a variable of long period--

four hundred and six days--whose changes of brilliance lie between magnitudes four and thirteen, but which exhibits much irregularity in its maxima. The other star, not named but easily recognized in the map, is sometimes called 17. It is an attractive double whose colors faintly reproduce those of beta. The magnitudes are five and eight, distance 26", p. 73? Where the two arms of the cross meet is gamma, whose remarkable cort 聞 e of small stars running in curved streams should not be missed. Use the lowest magnifying power.

At the extremity of the western arm of the cross is delta, a close double, difficult for telescopes of moderate aperture on account of the difference in the magnitudes of the components. We may succeed in dividing it with the five-inch. The magnitudes are three and eight, distance 1.5", p. 310? It is regarded as a binary of long and as yet unascertained period.

In omicron^2 we find a star of magnitude four and orange in color, having two blue companions, the first of magnitude seven and a half, distance 107", p. 174? and the second of magnitude five and a half, distance 358", p. 324? Farther north is psi, which presents to us the combination of a white five-and-a-half-magnitude star with a lilac star of magnitude seven and a half. The distance is 3", p. 184? A very pretty sight.

We now pass to the extremity of the other arm of the cross, near which lies the beautiful little double 49, whose components are of magnitudes six and eight, distance 2.8", p. 50? The colors are yellow and blue, conspicuous and finely contrasted. A neighboring double of similar hues is 52, in which the magnitudes are four and nine, distance 6", p. 60? Sweeping a little way northward we come upon an interesting binary, lambda, which is unfortunately beyond the dividing power of our largest glass. A good seven-inch or seven-and-a-half-inch should split it under favorable circumstances. Its magnitudes are six and seven, distance 0.66", p. 74?

The next step carries us to a very famous object, 61 Cygni, long known as the nearest star in the northern hemisphere of the heavens. It is a double which our three-inch will readily divide, the magnitudes being both six, distance 21", p. 122? The distance of 61 Cygni, according to Hall's parallax of 0.27", is about 70,000,000,000,000 miles. There is some question whether or not it is a binary, for, while the twin stars are both moving in the same direction in space with comparative rapidity, yet conclusive evidence of orbital motion is lacking. When one has noticed the contrast in apparent size between this comparatively near-by star, which the naked eye only detects with considerable difficulty, and some of its brilliant neighbors whose distance is so great as to be immeasurable with our present means, no better proof will be needed of the fact that the faintness of a star is not necessarily an indication of remoteness.

We may prepare our eyes for a beautiful exhibition of contrasted colors once more in the star . This is really a quadruple, although only two of its components are close and conspicuous. The magnitudes are five, six, seven and a half, and twelve; distances 2.4", p. 121? 208", p. 56? and 35", p. 264? The color of the largest star is white and that of its nearest companion blue; the star of magnitude seven and a half is also blue.

The star cluster 4681 is a fine sight with our largest glass. In the map we find the place marked where the new star of 1876 made its appearance. This was first noticed on November 24, 1876, when it shone with the brilliance of a star of magnitude three and a half. Its spectrum was carefully studied, especially by Vogel, and the very interesting changes that it underwent were noted. Within a year the star had faded to less than the tenth magnitude, and its spectrum had completely changed in appearance, and had come to bear a close resemblance to that of a planetary nebula. This has been quoted as a possible instance of a celestial collision through whose effects the solid colliding masses were vaporized and expanded into a nebula. At present the star is very faint and can only be seen with the most powerful telescopes. Compare with the case of Nova Aurig? previously discussed.

Underneath Cygnus we notice the small constellation Vulpecula. It contains a few objects worthy of attention, the first being the nebula 4532, the "dumb-bell nebula" of Lord Rosse. With the four-inch, and better with the five-inch, we are able to perceive that it consists of two close-lying tufts of misty light. Many stars surround it, and large telescopes show them scattered between the two main masses of the nebula. The Lick photographs show that its structure is spiral. The star 11 points out the place where a new star of the third magnitude appeared in 1670. Sigma 2695 is a close double, magnitudes six and eight, distance 0.96", p. 78?

We turn to map No. 18, and, beginning at the western end of the constellation Aquarius, we find the variable T, which ranges between magnitudes seven and thirteen in a period of about two hundred and three days. Its near neighbor Sigma 2729 is a very close double, beyond the separating power of our five-inch, the magnitudes being six and seven, distance 0.6", p. 176? Sigma 2745, also known as 12 Aquarii, is a good double for the three-inch. Its magnitudes are six and eight, distance 2.8", p. 190? In zeta we discover a beauty. It is a slow binary of magnitudes four and four, distance 3.1", p. 321? According to some observers both stars have a greenish tinge. The star 41 is a wider double, magnitudes six and eight, distance 5", p. 115? colors yellow and blue. The uncommon stellar contrast of white with light garnet is exhibited by tau, magnitudes six and nine, distance 27", p. 115? Yellow and blue occur again conspicuously in psi, magnitudes four and a half and eight and a half, distance 50", p. 310? Rose and emerald have been recorded as the colors exhibited in Sigma 2998, whose magnitudes are five and seven, distance 1.3", p. 346?

The variables S and R are both red. The former ranges between magnitudes eight and twelve, period two hundred and eighty days, and the latter between magnitudes six and eleven, period about three hundred and ninety days.

The nebula 4628 is Rosse's "Saturn nebula," so called because with his great telescope it presented the appearance of a nebulous model of the planet Saturn. With our five-inch we see it simply as a planetary nebula. We may also glance at another nebula, 4678, which appears circular and is pinned with a little star at the edge.

The small constellation Equuleus contains a surprisingly large number of interesting objects. Sigma 2735 is a rather close double, magnitudes six and eight, distance 1.8", p. 287? Sigma 2737 (the first star to the left of Sigma 2735, the name having accidentally been omitted from the map) is a beautiful triple, although the two closest stars, of magnitudes six and seven, can not be separated by our instruments. Their distance in 1886 was 0.78", p. 286? and they had then been closing rapidly since 1884, when the distance was 1.26". The third star, of magnitude eight, is distant 11", p. 75? Sigma 2744 consists of two stars, magnitudes six and seven, distance 1.4", p. 1.67? It is probably a binary. Sigma 2742 is wider double, magnitudes both six, distance 2.6", p. 225?

Another triple, one of whose components is beyond our reach, is gamma. Here the magnitudes are fifth, twelfth, and sixth, distances 2", p. 274?and 366". It would also be useless for us to try to separate delta, but it is interesting to remember that this is one of the closest of known double stars, the magnitudes being fourth and fifth, distance 0.4", p. 198? These data are from Hall's measurements in 1887. The star is, no doubt, a binary. With the five-inch we may detect one and perhaps two of the companion stars in the quadruple beta. The magnitudes are five, ten, and two eleven, distances 67", p. 309? 86", p. 276? and 6.5", p. 15? The close pair is comprised in the tenth-magnitude star.

Map No. 19 introduces us to the constellation Pegasus, which is comparatively barren to the naked eye, and by no means rich in telescopic phenomena. The star epsilon, of magnitude two and a half, has a blue companion of the eighth magnitude, distance 138", p. 324? colors yellow and violet. A curious experiment that may be tried with this star is described by Webb, who ascribes the discovery of the phenomenon to Sir John Herschel. When near the meridian the small star in epsilon appears, in the telescope, underneath the large one. If now the tube of the telescope be slightly swung from side to side the small star will appear to describe a pendulumlike movement with respect to the large one. The explanation suggested is that the comparative faintness of the small star causes its light to affect the retina of the eye less quickly than does that of its brighter companion, and, in consequence, the reversal of its apparent motion with the swinging of the telescope is not perceived so soon.

The third-magnitude star eta has a companion of magnitude ten and a half, distance 90", p. 340? The star beta, of the second magnitude, and reddish, is variable to the extent of half a magnitude in an irregular period, and gamma, of magnitude two and a half, has an eleventh-magnitude companion, distance 162", p. 285?

Our interest is revived on turning, with the guidance of map No. 20, from the comparative poverty of Pegasus to the spacious constellation Cetus. The first double star that we meet in this constellation is 26, whose components are of magnitudes six and nine, distance 16.4", p. 252? colors, topaz and lilac. Not far away is the closer double 42, composed of a sixth and a seventh magnitude star, distance 1.25", p. 350? The four-inch is capable of splitting this star, but we shall do better to use the five-inch. In passing we may glance at the tenth-magnitude companion to eta, distance 225", p. 304? Another wide pair is found in zeta, magnitudes three and nine, distance 185", p. 40?

The next step brings us to the wonderful variable omicron, or Mira, whose changes have been watched for three centuries, the first observer of the variability of the star having been David Fabricius in 1596. Not only is the range of variability very great, but the period is remarkably irregular. In the time of Hevelius, Mira was once invisible for four years. When brightest, the star is of about the second magnitude, and when faintest, of the ninth magnitude, but at maximum it seldom exhibits the greatest brilliance that it has on a few occasions shown itself capable of attaining. Ordinarily it begins to fade after reaching the fourth or fifth magnitude. The period averages about three hundred and thirty-one days, but is irregularly variable to the extent of twenty-five days. Its color is red, and its spectrum shows bright lines, which it is believed disappear when the star sinks to a minimum. Among the various theories proposed to account for such changes as these the most probable appears to be that which ascribes them to some cause analogous to that operating in the production of sun spots. The outburst of light, however, as pointed out by Scheiner, should be regarded as corresponding to the maximum and not the minimum stage of sun-spot activity. According to this view, the star is to be regarded as possessing an extensive atmosphere of hydrogen, which, during the maximum, is upheaved into enormous prominences, and the brilliance of the light from these prominences suffices to swamp the photospheric light, so that in the spectrum the hydrogen lines appear bright instead of dark.

It is not possible to suppose that Mira can be the center of a system of habitable planets, no matter what we may think of the more constant stars in that regard, because its radiation manifestly increases more than six hundred fold, and then falls off again to an equal extent once in every ten or eleven months. I have met people who can not believe that the Almighty would make a sun and then allow its energies "to go to waste," by not supplying it with a family of worlds. But I imagine that if they had to live within the precincts of Mira Ceti they would cry out for exemption from their own law of stellar utility.

The most beautiful double star in Cetus is gamma, magnitudes three and seven, distance 3", p. 288? hues, straw-color and blue. The leading star alpha, of magnitude two and a half, has a distant blue companion three magnitudes fainter, and between them are two minute stars, the southernmost of which is a double, magnitudes both eleven, distance 10", p. 225?

The variable S ranges between magnitudes seven and twelve in a somewhat irregular period of about eleven months, while R ranges between the seventh and the thirteenth magnitudes in a period of one hundred and sixty-seven days.

The constellation Eridanus, represented in map No. 21, contains a few fine double stars, one of the most interesting of which is 12, a rather close binary. The magnitudes are four and eight, distance 2", p. 327? We shall take the five-inch for this, and a steady atmosphere and sharp seeing will be necessary on account of the wide difference in the brightness of the component stars. Amateurs frequently fail to make due allowance for the effect of such difference. When the limit of separating power for a telescope of a particular aperture is set at 1" or 2", as the case may be, it is assumed that the stars composing the doubles on which the test is made shall be of nearly the same magnitude, or at least that they shall not differ by more than one or two magnitudes at the most. The stray light surrounding a comparatively bright star tends to conceal a faint companion, although the telescope may perfectly separate them so far as the stellar disks are concerned. Then, too, I have observed in my own experience that a very faint and close double is more difficult than a brighter pair not more widely separated, usually on account of the defect of light, and this is true even when the components of the faint double are of equal magnitude.

Sigma 470, otherwise known as 32 Eridani, is a superb object on account of the colors of its components, the larger star being a rich topaz and the smaller an ultramarine; while the difference in magnitude is not as great as in many of the colored doubles. The magnitudes are five and seven, distance 6.7", p. 348? The star gamma, of magnitude two and a half, has a tenth-magnitude companion, distant 51", p. 238? Sigma 516, also called 39 Eridani, consists of two stars of magnitudes six and nine, distance 6.4", p. 150? colors, yellow and blue. The supposed binary character of this star has not yet been established.

In omicron^2 we come upon an interesting triple star, two of whose components at any rate we can easily see. The largest component is of the fourth magnitude. At a distance of 82", p. 105? we find a tenth-magnitude companion. This companion is itself double, the magnitudes of its components being ten and eleven, distance 2.6", p. 98? Hall says of these stars that they "form a remarkable system." He has also observed a fourth star of the twelfth magnitude, distant 45" from the largest star, p. 85? This is apparently unconnected with the others, although it is only half as distant as the tenth-magnitude component is from the primary. Sigma 590 is interesting because of the similarity of its two components in size, both being of about the seventh magnitude, distance 10", p. 318?

Finally, we turn to the nebula 826. This is planetary in form and inconspicuous, but Lassell has described it as presenting a most extraordinary appearance with his great reflector--a circular nebula lying upon another fainter and larger nebula of a similar shape, and having a star in its center. Yet it may possibly be an immensely distant star cluster instead of a nebula, since its spectrum does not appear to be gaseous.

CHAPTER VII

PISCES, ARIES, TAURUS, AND THE NORTHERN STARS

"Now sing we stormy skies when Autumn weighs The year, and adds to nights and shortens days, And suns declining shine with feeble rays."-- DRYDEN'S VIRGIL.

The eastern end of Pisces, represented in map No. 22, includes most of the interesting telescopic objects that the constellation contains. We begin our exploration at the star numbered 55, a double that is very beautiful when viewed with the three-inch glass. The components are of magnitudes five and eight, distance 6.6", p. 192? The larger star is yellow and the smaller deep blue. The star 65, while lacking the peculiar charm of contrasted colors so finely displayed in 55, possesses an attraction in the equality of its components which are both of the sixth magnitude and milk-white. The distance is 4.5", p. 118? In 66 we find a swift binary whose components are at present far too close for any except the largest telescopes. The distance in 1894 was only 0.36", p. 329? The magnitudes are six and seven. In contrast with this excessively close double is psi, whose components are both of magnitude five and a half, distance 30", p. 160? Dropping down to 77 we come upon another very wide and pleasing double, magnitudes six and seven, distance 33", p. 82? colors white and lilac or pale blue. Hardly less beautiful is zeta magnitudes five and six, distance 24", p. 64? Finest of all is alpha, which exhibits a remarkable color contrast, the larger star being greenish and the smaller blue. The magnitudes are four and five, distance 3", p. 320? This star is a binary, but the motion is slow. The variable R ranges between magnitudes seven and thirteen, period three hundred and forty-four days.

The constellation Aries contains several beautiful doubles, all but one of which are easy for our smallest aperture. The most striking of these is gamma, which is historically interesting as the first double star discovered. The discovery was made by Robert Hooke in 1664 by accident, while he was following the comet of that year with his telescope. He expressed great surprise on noticing that the glass divided the star, and remarked that he had not met with a like instance in all the heavens. His observations could not have been very extensive or very carefully conducted, for there are many double stars much wider than gamma Arietis which Hooke could certainly have separated if he had examined them. The magnitudes of the components of gamma are four and four and a half, or, according to Hall, both four; distance 8.5", p. 180? A few degrees above gamma, passing by beta, is a wide double lambda, magnitudes five and eight, distance 37", p. 45? colors white and lilac or violet. Three stars are to be seen in 14: magnitudes five and a half, ten, and nine, distances 83", p. 36? and 106", p. 278? colors white, blue, and lilac. The star 30 is a very pretty double, magnitudes six and seven, distance 38.6", p. 273? Sigma 289 consists of a topaz star combined with a sapphire, magnitudes six and nine, distance 28.5", p. 0? The fourth-magnitude star 41 has several faint companions. The magnitudes of two of these are eleven and nine, distances 34", p. 203? and 130", p. 230? We discover another triple in pi, magnitudes five, eight, and eleven, distances 3.24", p. 122? and 25", p. 110? The double mentioned above as being too close for our three-inch glass is epsilon, which, however, can be divided with the four-inch, although the five-inch will serve us better. The magnitudes are five and a half and six, distance 1.26", p. 202? The star 52 has two companions, one of which is so close that our instruments can not separate it, while the other is too faint to be visible in the light of its brilliant neighbor without the aid of a very powerful telescope.

We are now about to enter one of the most magnificent regions in the sky, which is hardly less attractive to the naked eye than Orion, and which men must have admired from the beginning of their history on the earth, the constellation Taurus (map No. 23). Two groups of stars especially distinguish Taurus, the Hyades and the Pleiades, and both are exceedingly interesting when viewed with the lowest magnifying powers of our telescopes.

We shall begin with a little star just west of the Pleiades, Sigma 412, also called 7 Tauri. This is a triple, but we can see it only as a double, the third star being exceedingly close to the primary. The magnitudes are six and a half, seven, and ten, distances 0.3", p. 216? and 22", p. 62? In the Pleiades we naturally turn to the brightest star eta, or Alcyone, famous for having once been regarded as the central sun around which our sun and a multitude of other luminaries were supposed to revolve, and picturesque on account of the little triangle of small stars near it which the least telescopic assistance enables us to see. One may derive much pleasure from a study of the various groupings of stars in the Pleiades. Photography has demonstrated, what had long been suspected from occasional glimpses revealed by the telescope, that this celebrated cluster of stars is intermingled with curious forms of nebulae The nebulous matter appears in festoons, apparently attached to some of the larger stars, such as Alcyone, Merope, and Maia, and in long, narrow, straight lines, the most remarkable of which, a faintly luminous thread starting midway between Maia and Alcyone and running eastward some 40', is beaded with seven or eight stars. The width of this strange nebulous streak is, on an average, 3" or 4", and there is, perhaps, no more wonderful phenomenon anywhere in celestial space. Unfortunately, no telescope is able to show it, and all our knowledge about it is based upon photographs. It might be supposed that it was a nebulous disk seen edgewise, but for the fact that at the largest star involved in its course it bends sharply about 10?out of its former direction, and for the additional fact that it seems to take its origin from a curved offshoot of the intricate nebulous mass surrounding Maia. Exactly at the point where this curve is transformed into a straight line shines a small star! In view of all the facts the idea does not seem to be very far-fetched that in the Pleiades we behold an assemblage of suns, large and small, formed by the gradual condensation of a nebula, and in which evolution has gone on far beyond the stage represented by the Orion nebula, where also a group of stars may be in process of formation out of nebulous matter. If we look a little farther along this line of development, we may perceive in such a stellar assemblage as the cluster in Hercules, a still later phase wherein all the originally scattered material has, perhaps, been absorbed into the starry nuclei.

The yellow star Sigma 430 has two companions: magnitudes six, nine, and nine and a half, distances 26", p. 55? and 39", p. 302? The star 30 of the fifth magnitude has a companion of the ninth magnitude, distance 9", p. 58? colors emerald and purple, faint. An interesting variable, of the type of Algol, is lambda, which at maximum is of magnitude three and four tenths and at minimum of magnitude four and two tenths. Its period from one maximum to the next is about three days and twenty-three hours, but the actual changes occupy only about ten hours, and it loses light more swiftly than it regains it. A combination of red and blue is presented by Phi (mistakenly marked on map No. 23 as psi). The magnitudes are six and eight, distance 56", p. 242? A double of similar magnitudes is chi, distance 19", p. 25? Between the two stars which the naked eye sees in kappa is a minute pair, each of less than the eleventh magnitude, distance 5", p. 324? Another naked-eye double is formed by theta^1 and theta^2, in the Hyades. The magnitudes are five and five and a half, distance about 5' 37".

The leading star of Taurus, Aldebaran (alpha), is celebrated for its reddish color. The precise hue is rather uncertain, but Aldebaran is not orange as Betelgeuse in Orion is, and no correct eye can for an instant confuse the colors of these two stars, although many persons seem to be unable to detect the very plain difference between them in this respect. Aldebaran has been called "rose-red," and it would be an interesting occupation for an amateur to determine, with the aid of some proper color scale, the precise hue of this star, and of the many other stars which exhibit chromatic idiosyncrasy. Aldebaran is further interesting as being a standard first-magnitude star. With the four-inch glass we see without difficulty the tenth-magnitude companion following Aldebaran at a distance of 114", p. 35? There is an almost inexplicable charm about these faint attendants of bright stars, which is quite different from the interest attaching to a close and nearly equal pair. The impression of physical relationship is never lacking though it may be deceptive, and this awakens a lively appreciation of the vast differences of magnitude that exist among the different suns of space.

The actual size and might of this great red sun form an attractive subject for contemplation. As it appears to our eyes Aldebaran gives one twenty-five-thousand-millionth as much light as the sun, but if we were placed midway between them the star would outshine the sun in the ratio of not less than 160 to 1. And yet, gigantic as it is, Aldebaran is possibly a pygmy in comparison with Arcturus, whose possible dimensions were discussed in the chapter. Although Aldebaran is known to possess several of the metallic elements that exist in the sun, its spectrum differs widely from the solar spectrum in some respects, and more closely resembles that of Arcturus.

Other interesting objects in Taurus are sigma, divisible with the naked eye, magnitudes five and five and a half, distance 7'; Sigma 674, double, magnitudes six and nine, distance 10.5", p. 147? Sigma 716, double, magnitudes six and seven, distance 5", p. 200?-a pleasing sight; tau, triple, magnitudes four, ten and a half, and eleven, distances 36", p. 249? and 36", p. 60?-the ten-and-a-half-magnitude star is itself double, as discovered by Burnham; star cluster No. 1030, not quite as broad as the moon, and containing some stars as large as the eleventh magnitude; and nebula No. 1157, the so-called "Crab nebula" of Lord Rosse, which our glasses will show only as a misty patch of faint light, although large telescopes reveal in it a very curious structure.

We now turn to the cluster of circumpolar constellations sometimes called the Royal Family, in allusion to the well-known story of the Ethiopian king Cepheus and his queen Cassiopeia, whose daughter Andromeda was exposed on the seashore to be devoured by a monster, but who was saved by the hero Perseus. All these mythologic personages are represented in the constellations that we are about to study.[4] We begin with Andromeda (map No. 24). The leading star alpha marks one corner of the great square of Pegasus. The first star of telescopic interest that we find in Andromeda is , a double difficult on account of the faintness of the smaller component. The magnitudes are four and eleven, distance 49", p. 110? A few degrees north of the naked eye detects a glimmering point where lies the Great Nebula in Andromeda. This is indicated on the map by the number 116. With either of our three telescopes it is an interesting object, but of course it is advisable to use our largest glass in order to get as much light as possible. All that we can see is a long, shuttle-shaped nebulous object, having a brighter point near the center. Many stars are scattered over the field in its neighborhood, but the nebula itself, although its spectrum is peculiar in resembling that of a faint star, is evidently a gaseous or at any rate a meteoritic mass, since photographs show it to be composed of a series of imperfectly separated spirals surrounding a vast central condensation. This peculiarity of the Andromeda nebula, which is invisible with telescopes although conspicuous in the photographs, has, since its discovery a few years ago, given a great impetus to speculation concerning the transformation of nebulaeinto stars and star clusters. No one can look at a good photograph of this wonderful phenomenon without noticing its resemblance to the ideal state of things which, according to the nebular hypothesis, must once have existed in the solar system. It is to be remembered, however, that there is probably sufficient material in the Andromeda nebula to make a system many times, perhaps hundreds or thousands of times, as extensive as that of which our sun is the center. If one contemplates this nebula only long enough to get a clear perception of the fact that creation was not ended when, according to the Mosaic history, God, having in six days finished "the heavens and the earth and all the host of them," rested from all his work, a good blow will have been dealt for the cause of truth. Systems far vaster than ours are now in the bud, and long before they have bloomed, ambitious man, who once dreamed that all these things were created to serve him, will probably have vanished with the extinguishment of the little star whose radiant energy made his life and his achievements briefly possible.

[4] For further details on this subject see Astronomy with an Opera-glass.

In August, 1885, a new star of magnitude six and a half made its appearance suddenly near the center of the Andromeda nebula. Within one year it had disappeared, having gradually dwindled until the great Washington telescope, then the largest in use, no longer showed it. That this was a phenomenon connected with the nebula is most probable, but just what occurred to produce it nobody knows. The observed appearances might have been produced by a collision, and no better hypothesis has yet been suggested to account for them.

Near the opposite end of the constellation from alpha we find the most interesting of triple stars in gamma. The two larger components of this beautiful star are of magnitudes three and six, distance 10", colors golden yellow and deep blue. The three-inch shows them finely. The smaller star is itself double, its companion being of magnitude eight, distance when discovered in 1842 0.5", color bluish green. A few years ago this third star got so close to its primary that it was invisible even with the highest powers of the great Lick telescope, but at present it is widening again. In October, 1893, I had the pleasure of looking at gamma Andromed?with the Lick telescope, and at that time it was possible just to separate the third star. The angle seemed too small for certain measurement, but a single setting of the micrometer by Mr. Barnard, to whose kindness I was indebted for my view of the star, gave 0.17" as the approximate distance. In 1900 the distance had increased to 0.4", p. 115? The brilliance of color contrast between the two larger stars of gamma Andromed?is hardly inferior to that exhibited in beta Cygni, so that this star may be regarded as one of the most picturesque of stellar objects for small telescopes.

Other pleasing objects in this constellation are the binary star 36, magnitudes six and six and a half, distance 1", p. 17?-the two stars are slowly closing and the five-inch glass is required to separate them: the richly colored variable R, which fades from magnitude five and a half to invisibility, and then recovers its light in a period of about four hundred and five days; and the bright star cluster 457, which covers a space about equal to the area of the full moon.

Just south of the eastern end of Andromeda is the small constellation Triangulum, or the Triangles, containing two interesting objects. One of these is the beautiful little double 6, magnitudes five and six, distance 3.8", p. 77? colors yellow and blue; and the other, the nebula 352, which equals in extent the star cluster in Andromeda described above, but nevertheless appears very faint with our largest glass. Its faintness, however, is not an indication of insignificance, for to very powerful telescopes it exhibits a wonderful system of nuclei and spirals--another bit of chaos that is yielding by age-long steps to the influence of demiurgic forces.

A richer constellation than Andromeda, both for naked-eye and telescopic observation, is Perseus, which is especially remarkable for its star clusters. Two of these, 512 and 521, constitute the celebrated double cluster, sometimes called the Sword-hand of Perseus, and also chi Persei. To the smallest telescope this aggregation of stars, ranging in magnitude from six and a half to fourteen, and grouped about two neighboring centers, presents a marvelous appearance. As an educative object for those unaccustomed to celestial observations it may be compared among star clusters to beta Cygni among double stars, for the most indifferent spectator is struck with wonder in viewing it. All the other clusters in Perseus represented on the map are worth examining, although none of them calls for special mention, except perhaps 584, where we may distinguish at least a hundred separate stars within an area less than one quarter as expansive as the face of the moon.

Among the double stars of Perseus we note first eta, whose components are of magnitudes four and eight, distance 28", colors white and pale blue. The double epsilon is especially interesting on account of an alleged change of color from blue to red which the smaller star undergoes coincidently with a variation of brightness. The magnitudes are three and eight, distance 9", p. 9? An interesting multiple is zeta, two of whose stars at least we can see. The magnitudes are three, nine, ten, and ten, distances 13", p. 207? 90", and 112".

The chief attraction in Perseus is the changeful and wonderful beta, or Algol, the great typical star among the short-period variables. During the greater part of its period this star is of magnitude two and two tenths, but for a very short time, following a rapid loss of light, it remains at magnitude three and seven tenths. The difference, one magnitude and a half, corresponds to an actual difference in brightness in the ratio of 3.75 to 1. The entire loss of light during the declension occupies only four hours and a half. The star remains at its faintest for a few minutes only before a perceptible gain of light occurs, and the return to maximum is as rapid as was the preceding decline. The period from one minimum to the next is two days twenty hours forty-eight minutes fifty-three seconds, with an irregularity amounting to a few seconds in a year. The Arabs named the star Algol, or the Demon, on account of its eccentricity which did not escape their attention; and when Goodricke, in 1782, applied a scientific method of observation to it, the real cause of its variations was suggested by him, but his explanation failed of general acceptance until its truth was established by Prof. E. C. Pickering in 1880. This explanation gives us a wonderful insight into stellar constitution. According to it, Algol possesses a companion as large as the sun, but invisible, both because of its proximity to that star and because it yields no light, and revolving in a plane horizontal to our line of sight. The period of revolution is identical with the period of Algol's cycle of variation, and the diminution of light is caused by the interposition of the dark body as it sweeps along that part of its orbit lying between our point of view and the disk of Algol. In other words, once in every two days twenty hours and forty-nine minutes Algol, as seen from the earth, undergoes a partial eclipse.

In consequence of the great comparative mass of its dark companion, Algol itself moves in an orbit around their common center with a velocity quite sufficient to be detected by the shifting of the lines in its spectrum. By means of data thus obtained the mass, size, and distance apart of Algol and its singular comrade have been inferred. The diameter of Algol is believed to be about 1,125,000 miles, that of the dark body about 840,000 miles, and the mean distance from center to center 3,230,000 miles. The density of both the light and the dark star is slight compared with that of the sun, so that their combined mass is only two thirds as great as the sun's.

Mention has been made of a slight irregularity in Algol's period of variation. Basing his calculations upon this inequality, Dr. Chandler has put forward the hypothesis that there is another invisible body connected with Algol, and situated at a distance from it of about 1,800,000,000 miles, and that around this body, which is far more massive than the others, Algol and its companions revolve in a period of one hundred and thirty years! Dr. Chandler has earned the right to have his hypotheses regarded with respect, even when they are as extraordinary as that which has just been described. It needs no indulgence of the imagination to lend interest to Algol; the simple facts are sufficient. How did that bright star fall in with its black neighbors? Or were they created together?

Passing to the region covered by map No. 25, our eyes are caught by the curious figure, formed by the five brightest stars of the constellation Cassiopeia, somewhat resembling the letter W. Like Perseus, this is a rich constellation, both in star clusters and double stars. Among the latter we select as our first example sigma, in which we find a combination of color that is at once very unusual and very striking--green and blue. The magnitudes are five and seven, distance 3", p. 324? Another beautiful colored double is eta, whose magnitudes are four and seven and a half, distance 5", p. 200? colors white and purple. This is one of the comparatively small number of stars the measure of whose distance has been attempted, and a keen sense of the uncertainty of such measures is conveyed by the fact that authorities of apparently equal weight place eta Cassiopei?at such discordant distances as

124,000,000,000,000 miles, 70,000,000,000,000 miles, and 42,000,000,000,000 miles. It will be observed that the difference between the greatest and the least of these estimates is about double the entire distance given by the latter. The same thing is practically true of the various attempts to ascertain the distance of the other stars which have a perceptible parallax, even those which are evidently the nearest. In some cases the later measures increase the distance, in other cases they diminish it; in no case is there anything like a complete accord. Yet of course we are not to infer that it is hopeless to learn anything about the distances of the stars. With all their uncertainties and disagreements the few parallaxes we possess have laid a good foundation for a knowledge of the dimensions of at least the nearer parts of the universe.

We find an interesting triple in psi, the magnitudes of the larger components being four and a half and eight and a half, distance 30". The smaller star has a nine-and-a-half-magnitude companion, distance 3". A more beautiful triple is iota, magnitudes four, seven, and eight, distances 2", p. 256? and 7.5", p. 112? Cassiopeia contains many star clusters, three of which are indicated in the map. Of these 392 is perhaps the most interesting, as it includes stars of many magnitudes, among which are a red one of the eighth magnitude, and a ninth-magnitude double whose components are 8" apart. Not far from the star kappa we find the spot where the most brilliant temporary star on record made its appearance on November 11, 1572. Tycho Brahe studied this phenomenon during the entire period of its visibility, which lasted until March, 1574. It burst out suddenly with overpowering splendor, far outshining every fixed star, and even equaling Venus at her brightest. In a very short time it began to fade, regularly diminishing in brightness, and at the same time undergoing changes of color, ending in red, until it disappeared. It has never been seen since, and the suspicion once entertained that it was a variable with a period considerably exceeding three hundred years has not been confirmed. There is a tenth-magnitude star near the place given by Tycho as that occupied by the stranger. Many other faint stars are scattered about, however, and Tycho's measures were not sufficiently exact to enable us to identify the precise position of his star. If the phenomenon was due to a collision, no reappearance of the star is to be expected.

Camelopardalus is a very inconspicuous constellation, yet it furnishes considerable occupation for the telescope. Sigma 390, of magnitude five, has a companion of magnitude nine and a half, distance 15", 160? Sigma 385, also of the fifth magnitude, has a ninth-magnitude companion, distance only 2.4", p. 160? According to some observers, the larger star is yellow and the smaller white. The star 1 is a very pretty double, magnitudes both six, distance 10.4". Its neighbor 2 of magnitude six has an eighth-magnitude companion, distance 1.7", p. 278? The star 7 of magnitude five is also double, the companion of magnitude eight being distant only 1.2". A glance at star cluster 940, which shows a slight central condensation, completes our work in Camelopardalus, and we turn to Ursa Major, represented in map No. 26. Here there are many interesting doubles and triples. Beginning with iota we find at once occupation for our largest glass. The magnitudes are three and ten, distance 10", p. 357? In the double star 23 the magnitudes are four and nine, distance 23", p. 272? A more pleasing object is sigma^2, a greenish fifth-magnitude star which has an eighth-magnitude companion, distance 2.6", p. 245? A good double for our four-inch glass is xi, whose magnitudes are four and five, distance 1.87", p. 183? This is a binary with a period of revolution of about sixty years, and is interesting as the first binary star whose orbit was determined. Savary calculated it in 1828. Near by is nu, a difficult double, magnitudes four and ten and a half, distance 7", p. 147? In 57 we find again an easy double magnitudes six and eight, distance 5.5", p. 4? Another similar double is 65, magnitudes six and eight, distance 3.9", p. 38? A third star, magnitude seven, is seen at a distance of 114" from the primary.

We come now to Ursa Major's principal attraction zeta, frequently called Mizar. The naked eye perceives near it a smaller star, named Alcor. With the three-inch glass and a medium power we divide Mizar into two bright stars brilliantly contrasted in color, the larger being white and the smaller blue-green. Beside Alcor, several fainter stars are seen scattered over the field of view, and, taken all in all, there are very few equally beautiful sights in the starry heavens. The magnitudes of the double are three and four, distance 14.5", p. 148? The large star is again double, although no telescope has been able to show it so, its duplicity being revealed, like that of beta Aurig? by the periodical splitting of the lines in its spectrum.

Ursa Major contains several nebulaewhich may be glimpsed with telescopes of moderate dimensions. An interesting pair of these objects, both of which are included in one field of view, is formed by 1949 and 1950. The first named is the brighter of the two, its nucleus resembling a faint star. The nebula 2343 presents itself to us in the form of a faint, hazy star, but with large telescopes its appearance is very singular. According to a picture made by Lord Rosse, it bears no little resemblance to a skull, there being two symmetrically placed holes in it, each of which contains a star.

The portion of Canes Venatici, represented in map No. 26, contains two or three remarkable objects. Sigma 1606 is a close double, magnitudes six and seven, distance 1", p. 336? It is a pretty sight with the five-inch. The double star 2 is singular in that its larger component is red and its smaller blue; magnitudes six and eight, distance 11.4", p. 260? Still more beautiful is 12, commonly called Cor Caroli. This double is wide, and requires but a slight magnifying power. The magnitudes are three and six, distance 20", colors white or light yellow and blue. The nebula 3572, although we can see it only as a pair of misty specks, is in reality a very wonderful object. Lord Rosse's telescope has revealed in it a complicated spiral structure, recalling the photographs of the Andromeda nebula, and indicating that stupendous changes must be in process within it, although our records of observation are necessarily too brief to bring out any perceptible alteration of figure. It would seem that the astronomer has, of all men, the best reasons for complaining of the brevity of human life.

Lastly, we turn to Ursa Minor and the Pole Star. The latter is a celebrated double, not difficult, except with a telescope of less than three inches aperture in the hands of an inexperienced observer. The magnitudes are two and nine, distance 18.5". The small star has a dull blue color. In 1899 it was discovered by spectroscopic evidence that the Pole Star is triple. In pi' we see a wide double, magnitudes six and seven, distance 30", p. 83?

This completes our survey of the starry heavens.

CHAPTER VIII

SCENES ON THE PLANETS

"These starry globes far surpassed the earth in grandeur, and the latter looked so diminutive that our empire, which appeared only as a point on its surface, awoke my pity."--CICERO, THE DREAM OF SCIPIO.

Although amateurs have played a conspicuous part in telescopic discovery among the heavenly bodies, yet every owner of a small telescope should not expect to attach his name to a star. But he certainly can do something perhaps more useful to himself and his friends; he can follow the discoveries that others, with better appliances and opportunities, have made, and can thus impart to those discoveries that sense of reality which only comes from seeing things with one's own eyes. There are hundreds of things continually referred to in books and writings on astronomy which have but a misty and uncertain significance for the mere reader, but which he can easily verify for himself with the aid of a telescope of four or five inches aperture, and which, when actually confronted by the senses, assume a meaning, a beauty, and an importance that would otherwise entirely have escaped him. Henceforth every allusion to the objects he has seen is eloquent with intelligence and suggestion.

Take, for instance, the planets that have been the subject of so many observations and speculations of late years--Mars, Jupiter, Saturn, Venus. For the ordinary reader much that is said about them makes very little impression upon his mind, and is almost unintelligible. He reads of the "snow patches" on Mars, but unless he has actually seen the whitened poles of that planet he can form no clear image in his mind of what is meant. So the "belts of Jupiter" is a confusing and misleading phrase for almost everybody except the astronomer, and the rings of Saturn are beyond comprehension unless they have actually been seen.

It is true that pictures and photographs partially supply the place of observation, but by no means so successfully as many imagine. The most realistic drawings and the sharpest photographs in astronomy are those of the moon, yet I think nobody would maintain that any picture in existence is capable of imparting a really satisfactory visual impression of the appearance of the lunar globe. Nobody who has not seen the moon with a telescope--it need not be a large one--can form a correct and definite idea of what the moon is like.

The satisfaction of viewing with one's own eyes some of the things the astronomers write and talk about is very great, and the illumination that comes from such viewing is equally great. Just as in foreign travel the actual seeing of a famous city, a great gallery filled with masterpieces, or a battlefield where decisive issues have been fought out illuminates, for the traveler's mind, the events of history, the criticisms of artists, and the occurrences of contemporary life in foreign lands, so an acquaintance with the sights of the heavens gives a grasp on astronomical problems that can not be acquired in any other way. The person who has been in Rome, though he may be no archeologist, gets a far more vivid conception of a new discovery in the Forum than does the reader who has never seen the city of the Seven Hills; and the amateur who has looked at Jupiter with a telescope, though he may be no astronomer, finds that the announcement of some change among the wonderful belts of that cloudy planet has for him a meaning and an interest in which the ordinary reader can not share.

Jupiter is perhaps the easiest of all the planets for the amateur observer. A three-inch telescope gives beautiful views of the great planet, although a four-inch or a five-inch is of course better. But there is no necessity for going beyond six inches' aperture in any case. For myself, I should care for nothing better than my Byrne five-inch of fifty-two inches' focal distance. With such a glass more details are visible in the dark belts and along the bright equatorial girdle than can be correctly represented in a sketch before the rotation of the planet has altered their aspect, while the shadows of the satellites thrown upon the broad disk, and the satellites themselves when in transit, can be seen sometimes with exquisite clearness. The contrasting colors of various parts of the disk are also easily studied with a glass of four or five inches' aperture.

There is a charm about the great planet when he rides high in a clear evening sky, lording it over the fixed stars with his serene, unflickering luminousness, which no possessor of a telescope can resist. You turn the glass upon him and he floats into the field of view, with his cort 間 e of satellites, like a yellow-and-red moon, attended by four miniatures of itself. You instantly comprehend Jupiter's mastery over his satellites--their allegiance is evident. No one would for an instant mistake them for stars accidentally seen in the same field of view. Although it requires a very large telescope to magnify their disks to measurable dimensions, yet the smallest glass differentiates them at once from the fixed stars. There is something almost startling in their appearance of companionship with the huge planet-- this sudden verification to your eyes of the laws of gravitation and of central forces. It is easy, while looking at Jupiter amid his family, to understand the consternation of the churchmen when Galileo's telescope revealed that miniature of the solar system, and it is gratifying to gaze upon one of the first battle grounds whereon science gained a decisive victory for truth.

The swift changing of place among the satellites, as well as the rapidity of Jupiter's axial rotation, give the attraction of visible movement to the Jovian spectacle. The planet rotates in four or five minutes less than ten hours--in other words, it makes two turns and four tenths of a third turn while the earth is rolling once upon its axis. A point on Jupiter's equator moves about twenty-seven thousand miles, or considerably more than the entire circumference of the earth, in a single hour. The effect of this motion is clearly perceptible to the observer with a telescope on account of the diversified markings and colors of the moving disk, and to watch it is one of the greatest pleasures that the telescope affords.

It would be possible, when the planet is favorably situated, to witness an entire rotation of Jupiter in the course of one night, but the beginning and end of the observation would be more or less interfered with by the effects of low altitude, to say nothing of the tedium of so long a vigil. But by looking at the planet for an hour at a time in the course of a few nights every side of it will have been presented to view. Suppose the first observation is made between nine and ten o'clock on any night which may have been selected. Then on the following night between ten and eleven o'clock Jupiter will have made two and a half turns upon his axis, and the side diametrically opposite to that seen on the first night will be visible. On the third night between eleven and twelve o'clock Jupiter will have performed five complete rotations, and the side originally viewed will be visible again.

Owing to the rotundity of the planet, only the central part of the disk is sharply defined, and markings which can be easily seen when centrally located become indistinct or disappear altogether when near the limb. Approach to the edge of the disk also causes a foreshortening which sometimes entirely alters the aspect of a marking. It is advisable, therefore, to confine the attention mainly to the middle of the disk. As time passes, clearly defined markings on or between the cloudy belts will be seen to approach the western edge of the disk, gradually losing their distinctness and altering their appearance, while from the region of indistinct definition near the eastern edge other markings slowly emerge and advance toward the center, becoming sharper in outline and more clearly defined in color as they swing into view.

Watching these changes, the observer is carried away by the reflection that he actually sees the turning of another distant world upon its axis of rotation, just as he might view the revolving earth from a standpoint on the moon. Belts of reddish clouds, many thousands of miles across, are stretched along on each side of the equator of the great planet he is watching; the equatorial belt itself, brilliantly lemon-hued, or sometimes ruddy, is diversified with white globular and balloon-shaped masses, which almost recall the appearance of summer cloud domes hanging over a terrestrial landscape, while toward the poles shadowy expanses of gradually deepening blue or blue-gray suggest the comparative coolness of those regions which lie always under a low sun.

After a few nights' observation even the veriest amateur finds himself recognizing certain shapes or appearances--a narrow dark belt running slopingly across the equator from one of the main cloud zones to the other, or a rift in one of the colored bands, or a rotund white mass apparently floating above the equator, or a broad scallop in the edge of a belt like that near the site of the celebrated "red spot," whose changes of color and aspect since its first appearance in 1878, together with the light it has thrown on the constitution of Jupiter's disk, have all but created a new Jovian literature, so thoroughly and so frequently have they been discussed.

And, having noticed these recurring features, the observer will begin to note their relations to one another, and will thus be led to observe that some of them gradually drift apart, while others drift nearer; and after a time, without any aid from books or hints from observatories, he will discover for himself that there is a law governing the movements on Jupiter's disk. Upon the whole he will find that the swiftest motions are near the equator, and the slowest near the poles, although, if he is persistent and has a good eye and a good instrument, he will note exceptions to this rule, probably arising, as Professor Hough suggests, from differences of altitude in Jupiter's atmosphere. Finally, he will conclude that the colossal globe before him is, exteriorly at least, a vast ball of clouds and vapors, subject to tremendous vicissitudes, possibly intensely heated, and altogether different in its physical constitution, although made up of similar elements, from the earth. Then, if he chooses, he can sail off into the delightful cloud-land of astronomical speculation, and make of the striped and spotted sphere of Jove just such a world as may please his fancy--for a world of some kind it certainly is.

For many observers the satellites of Jupiter possess even greater attractions than the gigantic ball itself. As I have already remarked, their movements are very noticeable and lend a wonderful animation to the scene. Although they bear classical names, they are almost universally referred to by their Roman numbers, beginning with the innermost, whose symbol is I, and running outward in regular order II, III, and IV.[5] The minute satellite much nearer to the planet than any of the others, which Mr. Barnard discovered with the Lick telescope in 1892, is called the fifth, although in the order of distance it would be the first. In size and importance, however, it can not rank with its comparatively gigantic brothers. Of course, no amateur's telescope can afford the faintest glimpse of it.

[5] Their names, in the same order as their numbers, are Io, Europa, Ganymede, and Callisto.

Satellite I, situated at a mean distance of 261,000 miles from Jupiter's center--about 22,000 miles farther than the moon is from the earth--is urged by its master's overpowering attraction to a speed of 320 miles per minute, so that it performs a complete revolution in about forty-two hours and a half. The others, of course, move more slowly, but even the most distant performs its revolution in several hours less than sixteen days. The plane of their orbits is presented edgewise toward the earth, from which it follows that they appear to move back and forth nearly in straight lines, some apparently approaching the planet, while others are receding from it. The changes in their relative positions, which can be detected from hour to hour, are very striking night after night, and lead to a great variety of arrangements always pleasing to the eye.

The most interesting phenomena that they present are their transits and those of their round, black shadows across the face of the planet; their eclipses by the planet's shadow, when they disappear and afterward reappear with astonishing suddenness; and their occultations by the globe of Jupiter. Upon the whole, the most interesting thing for the amateur to watch is the passage of the shadows across Jupiter. The distinctness with which they can be seen when the air is steady is likely to surprise, as it is certain to delight, the observer. When it falls upon a light part of the disk the shadow of a satellite is as black and sharply outlined as a drop of ink; on a dark-colored belt it can not so easily be seen.

It is more difficult to see the satellites themselves in transit. There appears to be some difference among them as to visibility in such circumstances. Owing to their luminosity they are best seen when they have a dark belt for a background, and are least easily visible when they appear against a bright portion of the planet. Every observer should provide himself with a copy of the American Ephemeris for the current year, wherein he will find all the information needed to enable him to identify the various satellites and to predict, by turning Washington mean time into his own local time, the various phenomena of the transits and eclipses.

While a faithful study of the phenomena of Jupiter is likely to lead the student to the conclusion that the greatest planet in our system is not a suitable abode for life, yet the problem of its future, always fascinating to the imagination, is open; and whosoever may be disposed to record his observations in a systematic manner may at least hope to render aid in the solution of that problem.

Saturn ranks next to Jupiter in attractiveness for the observer with a telescope. The rings are almost as mystifying to-day as they were in the time of Herschel. There is probably no single telescopic view that can compare in the power to excite wonder with that of Saturn when the ring system is not so widely opened but that both poles of the planet project beyond it. One returns to it again and again with unflagging interest, and the beauty of the spectacle quite matches its singularity. When Saturn is in view the owner of a telescope may become a recruiting officer for astronomy by simply inviting his friends to gaze at the wonderful planet. The silvery color of the ball, delicately chased with half-visible shadings, merging one into another from the bright equatorial band to the bluish polar caps; the grand arch of the rings, sweeping across the planet with a perceptible edging of shadow; their sudden disappearance close to the margin of the ball, where they go behind it and fall straightway into night; the manifest contrast of brightness, if not of color, between the two principal rings; the fine curve of the black line marking the

1,600-mile gap between their edges--these are some of the elements of a picture that can never fade from the memory of any one who has once beheld it in its full glory.

Saturn's moons are by no means so interesting to watch as are those of Jupiter. Even the effect of their surprising number (raised to nine by Professor Pickering's discovery in 1899 of a new one which is almost at the limit of visibility, and was found only with the aid of photography) is lost, because most of them are too faint to be seen with ordinary telescopes, or, if seen, to make any notable impression upon the eye. The two largest--Titan and Japetus--are easily found, and Titan is conspicuous, but they give none of that sense of companionship and obedience to a central authority which strikes even the careless observer of Jupiter's system. This is owing partly to their more deliberate movements and partly to the inclination of the plane of their orbits, which seldom lies edgewise toward the earth.

But the charm of the peerless rings is abiding, and the interest of the spectator is heightened by recalling what science has recently established as to their composition. It is marvelous to think, while looking upon their broad, level surfaces--as smooth, apparently, as polished steel, though thirty thousand miles across--that they are in reality vast circling currents of meteoritic particles or dust, through which run immense waves, condensation and rarefaction succeeding one another as in the undulations of sound. Yet, with all their inferential tumult, they may actually be as soundless as the depths of interstellar space, for Struve has shown that those spectacular rings possess no appreciable mass, and, viewed from Saturn itself, their (to us) gorgeous seeming bow may appear only as a wreath of shimmering vapor spanning the sky and paled by the rivalry of the brighter stars.

In view of the theory of tidal action disrupting a satellite within a critical distance from the center of its primary, the thoughtful observer of Saturn will find himself wondering what may have been the origin of the rings. The critical distance referred to, and which is known as Roche's limit, lies, according to the most trustworthy estimates, just outside the outermost edge of the rings. It follows that if the matter composing the rings were collected into a single body that body would inevitably be torn to pieces and scattered into rings; and so, too, if instead of one there were several or many bodies of considerable size occupying the place of the rings, all of these bodies would be disrupted and scattered. If one of the present moons of Saturn--for instance, Mimas, the innermost hitherto discovered--should wander within the magic circle of Roche's limit it would suffer a similar fate, and its particles would be disseminated among the rings. One can hardly help wondering whether the rings have originated from the demolition of satellites--Saturn devouring his children, as the ancient myths represent, and encircling himself, amid the fury of destruction, with the dust of his disintegrated victims. At any rate, the amateur student of Saturn will find in the revelations of his telescope the inspirations of poetry as well as those of science, and the bent of his mind will determine which he shall follow.

Professor Pickering's discovery of a ninth satellite of Saturn, situated at the great distance of nearly eight million miles from the planet, serves to call attention to the vastness of the "sphere of activity" over which the ringed planet reigns. Surprising as the distance of the new satellite appears when compared with that of our moon, it is yet far from the limit where Saturn's control ceases and that of the sun becomes predominant. That limit, according to Prof. Asaph Hall's calculation, is nearly 30,000,000 miles from Saturn's center, while if our moon were removed to a distance a little exceeding 500,000 miles the earth would be in danger of losing its satellite through the elopement of Artemis with Apollo.

Although, as already remarked, the satellites of Saturn are not especially interesting to the amateur telescopist, yet it may be well to mention that, in addition to Titan and Japetus, the satellite named Rhea, the fifth in order of distance from the planet, is not a difficult object for a three-or four-inch telescope, and two others considerably fainter than Rhea--Dione (the fourth) and Tethys (the third)--may be seen in favorable circumstances. The others-- Mimas (the first), Enceladus (the second), and Hyperion (the seventh)--are beyond the reach of all but large telescopes. The ninth satellite, which has received the name of Ph[oe]be, is much fainter than any of the others, its stellar magnitude being reckoned by its discoverer at about 15.5.

Mars, the best advertised of all the planets, is nearly the least satisfactory to look at except during a favorable opposition, like those of 1877 and 1892, when its comparative nearness to the earth renders some of its characteristic features visible in a small telescope. The next favorable opposition will occur in 1907.

When well seen with an ordinary telescope, say a four-or five-inch glass, Mars shows three peculiarities that may be called fairly conspicuous--viz., its white polar cap, its general reddish, or orange-yellow, hue, and its dark markings, one of the clearest of which is the so-called Syrtis Major, or, as it was once named on account of its shape, "Hourglass Sea." Other dark expanses in the southern hemisphere are not difficult to be seen, although their outlines are more or less misty and indistinct. The gradual diminution of the polar cap, which certainly behaves in this respect as a mass of snow and ice would do, is a most interesting spectacle. As summer advances in the southern hemisphere of Mars, the white circular patch surrounding the pole becomes smaller, night after night, until it sometimes disappears entirely even from the ken of the largest telescopes. At the same time the dark expanses become more distinct, as if the melting of the polar snows had supplied them with a greater depth of water, or the advance of the season had darkened them with a heavier growth of vegetation.

The phenomena mentioned above are about all that a small telescope will reveal. Occasionally a dark streak, which large instruments show is connected with the mysterious system of "canals," can be detected, but the "canals" themselves are far beyond the reach of any telescope except a few of the giants handled by experienced observers. The conviction which seems to have forced its way into the minds even of some conservative astronomers, that on Mars the conditions, to use the expression of Professor Young, "are more nearly earthlike than on any other of the heavenly bodies which we can see with our present telescopes," is sufficient to make the planet a center of undying interest notwithstanding the difficulties with which the amateur is confronted in his endeavors to see the details of its markings.

THE ILLUMINATION OF VENUS'S ATMOSPHERE AT THE BEGINNING OF HER TRANSIT ACROSS THE SUN.

In Venus "the fatal gift of beauty" may be said, as far as our observations are concerned, to be matched by the equally fatal gift of brilliance. Whether it be due to atmospheric reflection alone or to the prevalence of clouds, Venus is so bright that considerable doubt exists as to the actual visibility of any permanent markings on her surface. The detailed representations of the disk of Venus by Mr. Percival Lowell, showing in some respects a resemblance to the stripings of Mars, can not yet be accepted as decisive. More experienced astronomers than Mr. Lowell have been unable to see at all things which he draws with a fearless and unhesitating pencil. That there are some shadowy features of the planet's surface to be seen in favourable circumstances is probable, but the time for drawing a "map of Venus" has not yet come.

The previous work of Schiaparelli lends a certain degree of probability to Mr. Lowell's observations on the rotation of Venus. This rotation, according to the original announcement of Schiaparelli, is probably performed in the same period as the revolution around the sun. In other words, Venus, if Schiaparelli and Lowell are right, always presents the same side to the sun, possessing, in consequence, a day hemisphere and a night hemisphere which never interchange places. This condition is so antagonistic to all our ideas of what constitutes habitability for a planet that one hesitates to accept it as proved, and almost hopes that it may turn out to have no real existence. Venus, as the twin of the earth in size, is a planet which the imagination, warmed by its sunny aspect, would fain people with intelligent beings a little fairer than ourselves; but how can such ideas be reconciled with the picture of a world one half of which is subjected to the merciless rays of a never-setting sun, while the other half is buried in the fearful gloom and icy chill of unending night?

Any amateur observer who wishes to test his eyesight and his telescope in the search of shades or markings on the disk of Venus by the aid of which the question of its rotation may finally be settled should do his work while the sun is still above the horizon. Schiaparelli adopted that plan years ago, and others have followed him with advantage. The diffused light of day serves to take off the glare which is so serious an obstacle to the successful observation of Venus when seen against a dark sky. Knowing the location of Venus in the sky, which can be ascertained from the Ephemeris, the observer can find it by day. If his telescope is not permanently mounted and provided with "circles" this may not prove an easy thing to do, yet a little perseverance and ingenuity will effect it. One way is to find, with a star chart, some star whose declination is the same, or very nearly the same, as that of Venus, and which crosses the meridian say twelve hours ahead of her. Then set the telescope upon that star, when it is on the meridian at night, and leave it there, and the next day, twelve hours after the star crossed the meridian, look into your telescope and you will see Venus, or, if not, a slight motion of the tube will bring her into view.

For many amateurs the phases of Venus will alone supply sufficient interest for telescopic observation. The changes in her form, from that of a round full moon when she is near superior conjunction to the gibbous, and finally the half-moon phase as she approaches her eastern elongation, followed by the gradually narrowing and lengthening crescent, until she is a mere silver sickle between the sun and the earth, form a succession of delightful pictures.

Not very much can be said for Mercury as a telescopic object. The little planet presents phases like those of Venus, and, according to Schiaparelli and Lowell, it resembles Venus in its rotation, keeping always the same side to the sun. In fact, Schiaparelli's discovery of this peculiarity in the case of Mercury preceded the similar discovery in the case of Venus. There are markings on Mercury which have reminded some astronomers of the moon, and there are reasons for thinking that the planet can not be a suitable abode for living beings, at least for beings resembling the inhabitants of the earth.

Uranus and Neptune are too far away to present any attraction for amateur observers.

CHAPTER IX

THE MOUNTAINS AND PLAINS OF THE MOON, AND THE SPECTACLES OF THE SUN

"... the Moon, whose orb The Tuscan artist views through optic glass At evening from the top of Fesol? Or in Valdarno, to descry new lands, Rivers or mountains in her spotty globe."--PARADISE LOST.

The moon is probably the most interesting of all telescopic objects. This arises from its comparative nearness to the earth. A telescope magnifying 1,000 diameters brings the moon within an apparent distance of less than 240 miles. If telescopes are ever made with a magnifying power of 10,000 diameters, then, provided that atmospheric difficulties can be overcome, we shall see the moon as if it were only about twenty miles off, and a sensitive astronomer might be imagined to feel a little hesitation about gazing so closely at the moon--as if he were peering into a neighbor world's window.

But a great telescope and a high magnifying power are not required to interest the amateur astronomer in the study of the moon. Our three-inch telescope is amply sufficient to furnish us with entertainment for many an evening while the moon is running through its phases, and we shall find delight in frequently changing the magnifying power as we watch the lunar landscapes, because every change will present them in a different aspect.

It should be remembered that a telescope, unless a terrestrial eyepiece or prism is employed, reverses such an object as the moon top for bottom. Accordingly, if the moon is on or near the meridian when the observations are made, we shall see the north polar region at the bottom and the south polar region at the top. In other words, the face of the moon as presented in the telescope will be upside down, north and south interchanging places as compared with their positions in a geographical map. But east and west remain unaltered in position, as compared with such a map--i. e., the eastern hemisphere of the moon is seen on the right and the western hemisphere on the left. It is the moon's western edge that catches the first sunlight when "new moon" begins, and, as the phase increases, passing into "first quarter" and from that to "full moon," the illumination sweeps across the disk from west to east.

The narrow sickle of the new moon, hanging above the sunset, is a charming telescopic sight. Use a low power, and observe the contrast between the bright, smooth round of the sunward edge, which has almost the polish of a golden rim, and the irregular and delicately shaded inner curve, where the adjacent mountains and plains picturesquely reflect or subdue the sunshine. While the crescent grows broader new objects are continually coming into view as the sun rises upon them, until at length one of the most conspicuous and remarkable of the lunar "seas," the Mare Crisium, or Sea of Crises, lies fully displayed amid its encircling peaks, precipices, and craters. The Mare Crisium is all in the sunlight between the third and fourth day after "new moon." It is about 350 by 280 miles in extent, and if ever filled with water must have been a very deep sea, since its arid bed lies at a great but not precisely ascertained depth below the general level of the moon. There are a few small craters on the floor of the Mare Crisium, the largest bearing the name of Picard, and its borders are rugged with mountains. On the southwestern side is a lofty promontory, 11,000 feet in height, called Cape Agarum. At the middle of the eastern side a kind of bay opens deep in the mountains, whose range here becomes very narrow. Southeast of this bay lies a conspicuous bright point, the crater mountain Proclus, on which the sun has fully risen in the fourth day of the moon, and which reflects the light with extraordinary liveliness. Adjoining Proclus on the east and south is a curious, lozenge-shaped flat, broken with short, low ridges, and possessing a most peculiar light-brown tint, easily distinguished from the general color tone of the lunar landscapes. It would be interesting to know what was passing in the mind of the old astronomer who named this singular region Palus Somnii. It is not the only spot on the moon which has been called a "marsh," and to which an unexplained connection with dreams has been ascribed.

Nearly on the same meridian with Proclus, at a distance of about a hundred miles northward, lies a fine example of a ring mountain, rather more than forty miles in diameter, and with peak-tipped walls which in some places are 13,000 feet in height, as measured from the floor within. This is Macrobius. There is an inconspicuous central mountain in the ring.

North of the Mare Crisium, and northwest of Macrobius, we find a much larger mountain ring, oblong in shape and nearly eighty miles in its greatest diameter. It is named Cleomenes. The highest point on its wall is about 10,000 feet above the interior. Near the northeast corner of the wall yawns a huge and very deep crater, Tralles, while at the northern end is another oblong crater mountain called Burckhardt.

From Cleomenes northward to the pole, or to the northern extremity of the crescent, if our observations are made during new moon, the ground appears broken with an immense number of ridges, craters, and mountain rings, among which we may telescopically wander at will. One of the more remarkable of these objects, which may be identified with the aid of Lunar Chart No. 1, is the vast ringed plain near the edge of the disk, named Gauss. It is more than a hundred and ten miles in diameter. Owing to its situation, so far down the side of the lunar globe, it is foreshortened into a long ellipse, although in reality it is nearly a circle. A chain of mountains runs north and south across the interior plain. Geminus, Berzelius, and Messala are other rings well worth looking at. The remarkable pair called Atlas and Hercules demand more than passing attention. The former is fifty-five and the latter forty-six miles in diameter. Each sinks 11,000 feet below the summit of the loftiest peak on its encircling wall. Both are full of interesting detail sufficient to occupy the careful observer for many nights. The broad ring bearing the name of Endymion is nearly eighty miles in diameter, and has one peak 15,000 feet high. The interior plain is flat and dark. Beyond Endymion on the edge of the disk is part of a gloomy plain called the Mare Humboltianum.

After glancing at the crater-shaped mountains on the western and southern border of the Mare Crisium, Alhazen, Hansen, Condorcet, Firmicus, etc., we pass southward into the area covered in Lunar Chart No. 2. The long dark plain south of the Mare Crisium is the Mare Fecunditatis, though why it should have been supposed to be particularly fecund, or fertile, is by no means clear. On the western border of this plain, about three hundred miles from the southern end of the Mare Crisium, is the mountain ring, or circumvallation, called Langrenus, about ninety miles across and in places 10,000 feet high. There is a fine central mountain with a number of peaks. Nearly a hundred miles farther south, on the same meridian, lies an equally extensive mountain ring named Vendelinus. The broken and complicated appearance of its northern walls will command the observer's attention. Another similar step southward, and still on the same meridian brings us to a yet finer mountain ring, slightly larger than the others, and still more complicated in its walls, peaks, and terraces, and in its surroundings of craters, gorges, and broken ridges. This is Petavius. West of Petavius, on the very edge of the disk, is a wonderful formation, a walled plain named Humboldt, which is looked down upon at one point near its eastern edge by a peak 16,000 feet in height. About a hundred and forty miles south of Petavius is the fourth great mountain ring lying on the same meridian. Its name is Furnerius. Look particularly at the brilliantly shining crater on the northeast slope of the outer wall of Furnerius.

Suppose that our observations are now interrupted, to be resumed when the moon, about "seven days old," is in its first quarter. If we had time, it would be a most interesting thing to watch the advance of the lunar sunrise every night, for new beauties are displayed almost from hour to hour; but, for the purposes of our description it is necessary to curtail the observations. At first quarter one half of the lunar hemisphere which faces the earth is illuminated by the sun, and the line of sunrise runs across some of the most wonderful regions of the moon.

We begin, referring once more to Lunar Chart No. 1, in the neighborhood of the north pole of the moon. Here the line along which day and night meet is twisted and broken, owing to the roughness of the lunar surface. About fifteen degrees southwest of the pole lies a remarkable square-cornered, mountain-bordered plain, about forty miles in length, called Barrow. Very close to the pole is a ring mountain, about twenty-five miles in diameter, whose two loftiest peaks, 8,000 to 9,000 feet high, according to Neison, must, from their situation, enjoy perpetual day.

The long, narrow, dark plain, whose nearest edge is about thirty degrees south of the pole, is the Mare Frigoris, bordered on both sides by uplands and mountains. At its southern edge we find the magnificent Aristoteles, a mountain ring, sixty miles across, whose immense wall is composed of terraces and ridges running up to lofty peaks, which rise nearly 11,000 feet above the floor of the valley. About a hundred miles south of Aristoteles is Eudoxus, another fine mountain ring, forty miles in diameter, and quite as deep as its northern neighbor. These two make a most striking spectacle.

We are now in the neighborhood of the greatest mountain chains on the moon, the lunar Alps lying to the east and the lunar Caucasus to the south of Aristoteles and Eudoxus, while still farther south, separated from the Caucasus by a strait not more than a hundred miles broad, begins the mighty range of the lunar Apennines. We first turn the telescope on the Alps. As the line of sunrise runs directly across their highest peaks, the effect is startling. The greatest elevations are about 12,000 feet. The observer's eye is instantly caught by a great valley, running like a furrow through the center of the mountain mass, and about eighty or ninety miles in length. The sealike expanse south and southeast of the Alps is the Mare Imbrium, and it is along the coast of this so-called sea that the Alps attain their greatest height. The valley, or gorge, above mentioned, appears to cut through the loftiest mountains and to reach the "coast," although it is so narrowed and broken among the greater peaks that its southern portion is almost lost before it actually reaches the Mare Imbrium. Opening wider again as it enters the Mare, it forms a deep bay among precipitous mountains.

The Caucasus Mountains are not so lofty nor so precipitous as the Alps, and consequently have less attraction for the observer. They border the dark, oval plain of the Mare Serenitatis on its northeastern side. The great bay running out from the Mare toward the northwest, between the Caucasus and the huge mountain ring of Posidonius, bears the fanciful name of Lacus Somniorum. In the old days when the moon was supposed to be inhabited, those terrestrial godfathers, led by the astronomer Riccioli, who were busy bestowing names upon the "seas" and mountains of our patient satellite, may have pleased their imagination by picturing this arm of the "Serene Sea" as a peculiarly romantic sheet of water, amid whose magical influences the lunar gentlefolk, drifting softly in their silver galleons and barges, and enjoying the splendors of "full earth" poured upon their delightful little world, were accustomed to fall into charming reveries, as even we hard-headed sons of Adam occasionally do when the waters under the keel are calm and smooth and the balmy air of a moonlit night invokes the twin spirits of poetry and music.

Posidonius, the dominating feature of the shore line here, is an extraordinary example of the many formations on the moon which are so different from everything on the earth that astronomers do not find it easy to bestow upon them names that truly describe them. It may be called a ring mountain or a ringed plain, for it is both. Its diameter exceeds sixty miles, and the interior plain lies about 2,000 feet below the outer surface of the lunar ground. The mountain wall surrounding the ring is by no means remarkable for elevation, its greatest height not exceeding 6,000 feet, but, owing to the broad sweep of the curved walls, the brightness of the plain they inclose, and the picturesque irregularity of the silhouette of shadow thrown upon the valley floor by the peaks encircling it, the effect produced upon the observer is very striking and attractive.

Having finished with Posidonius and glanced across the broken region of the Taurus Mountains toward the west, we turn next to consider the Mare Serenitatis. This broad gray plain, which, with a slight magnifying power, certainly looks enough like a sea to justify the first telescopists in thinking that it might contain water, is about 430 by 425 miles in extent, its area being 125,000 square miles. Running directly through its middle, nearly in a north and south line, is a light streak, which even a good opera glass shows. This streak is the largest and most wonderful of the many similar rays which extend on all sides from the great crater, or ring, of Tycho in the southern hemisphere. The ray in question is more than 2,000 miles long, and, like its shorter congeners, it turns aside for nothing; neither "sea," nor peak, nor mountain range, nor crater ring, nor gorge, nor canyon, is able to divert it from its course. It ascends all heights and drops into all depths with perfect indifference, but its continuity is not broken. When the sun does not illuminate it at a proper angle, however, the mysterious ray vanishes. Is it a metallic vein, or is it volcanic lava or ash? Was the globe of the moon once split open along this line?

The Mare Serenitatis is encircled by mountain ranges to a greater extent than any of the other lunar "seas." On its eastern side the Caucasus and the Apennines shut it in, except for a strait a hundred miles broad, by means of which it is connected with the Mare Imbrium. On the south the range of the Hetus Mountains borders it, on the north and northwest the Caucasus and the Taurus Mountains confine it, while on the west, where again it connects itself by a narrow strait with another "sea," the Mare Tranquilitatis, it encounters the massive uplift of Mount Argos. Not far from the eastern strait is found the remarkable little crater named Linn? not conspicuous on the gray floor of the Mare, yet easily enough found, and very interesting because a considerable change of form seems to have come over this crater some time near the middle of the nineteenth century. In referring to it as a crater it must not be forgotten that it does not form an opening in the top of a mountain. In fact, the so-called craters on the moon, generally speaking, are simply cavities in the lunar surface, whose bottoms lie deep below the general level, instead of being elevated on the summit of mountains, and inclosed in a conical peak. In regard to the alleged change in Linn? it has been suggested, not that a volcanic eruption brought it about, but that a downfall of steep walls, or of an unsupported rocky floor, was the cause. The possibility of such an occurrence, it must be admitted, adds to the interest of the observer who regularly studies the moon with a telescope.

Just on the southern border of the Mare, the beautiful ring Menelaus lies in the center of the chain of the Mountains. The ring is about twenty miles across, and its central peak is composed of some highly reflecting material, so that it shines very bright. The streak or ray from Tycho which crosses the Mare Serenitatis passes through the walls of Menelaus, and perhaps the central peak is composed of the same substance that forms the ray. Something more than a hundred miles east-southeast from Menelaus, in the midst of the dark Mare Vaporum, is another brilliant ring mountain which catches the eye, Manilius. It exceeds Menelaus in brightness as well as in size, its diameter being about twenty-five miles. There is something singular underlying the dark lunar surface here, for not only is Manilius extraordinarily brilliant in contrast with the surrounding plain, but out of that plain, about forty miles toward the east, projects a small mountain which is also remarkable for its reflecting properties, as if the gray ground were underlain by a stratum of some material that flashes back the sunlight wherever it is exposed. The crater mountain, Sulpicius Gallus, on the border of the Mare, north of Manilius and east of Menelaus, is another example of the strange shining quality of certain formations on the moon.

Follow next the range westward until the attention falls upon the great ring mountain Plinius, more than thirty miles across, and bearing an unusual resemblance to a fortification. Mr. T. G. Elger, the celebrated English selenographer, says of Plinius that, at sunrise, "it reminds one of a great fortress or redoubt erected to command the passage between the Mare Tranquilitatis and the Mare Serenitatis." But, of course, the resemblance is purely fanciful. Men, even though they dwelt in the moon, would not build a rampart 6,000 feet high!

Mount Argos, at the southwest corner of the Mare Serenitatis, is a very wonderful object when the sun has just risen upon it. This occurs five days after the new moon.

Returning to the eastern extremity of the Mare, we glance, in passing, at the precipitous Mount Hadley, which rises more than 15,000 feet above the level of the Mare and forms the northern point of the Apennine range. Passing into the region of the Mare Imbrium, whose western end is divided into the Palus Putredinis on the south and the Palus Nebularum on the north, we notice three conspicuous ring mountains, Cassini near the Alps, and Aristillus and Autolycus, a beautiful pair, nearly opposite the strait connecting the two Maria. Cassini is thirty-six miles in diameter, Aristillus thirty-four, and Autolycus twenty-three. The first named is shallow, only 4,000 feet in depth from the highest point of its wall, while Aristillus carries some peaks on its girdle 11,000 feet high. Autolycus, like Cassini, is of no very great depth.

Westward from the middle of an imaginary line joining Aristillus and Cassini is the much smaller crater Thetus. Outside the walls of this are a number of craterlets, and a French astronomer, Charbonneaux, of the Meudon Observatory, reported in December, 1900, that he had repeatedly observed white clouds appearing and disappearing over one of these small craters.

South of the Mare Vaporum are found some of the most notable of those strange lunar features that are called "clefts" or "rills." Two crater mountains, in particular, are connected with them, Ariades at the eastern edge of the Mare Tranquilitatis and Hyginus on the southern border of the Mare Vaporum. These clefts appear to be broad and deep chasms, like the canyons cut by terrestrial rivers, but it can not be believed that the lunar canyons are the work of rivers. They are rather cracks in the lunar crust, although their bottoms are frequently visible. The principal cleft from Ariades runs eastward and passes between two neighboring craters, the southern of which is named Silberschlag, and is noteworthy for its brightness. The Hyginus cleft is broader and runs directly through the crater ring of that name.

The observer will find much to interest him in the great, irregular, and much-broken mountain ring called Julius Caesar, as well as in the ring mountains, Godin, Agrippa, and Triesnecker. The last named, besides presenting magnificent shadows when the sunlight falls aslant upon it, is the center of a complicated system of rills, some of which can be traced with our five-inch glass.

We next take up Lunar Chart No. 2, and pay a telescopic visit to the southwestern quarter of the lunar world. The Mare Tranquilitatismerges through straits into two southern extensions, the Mare Fecunditatis and the Mare Nectaris. The great ring mountains or ringed plains, Langrenus, Vendelinus, Petavius, and Furnerius, all lying significantly along the same lunar meridian, have already been noticed. Their linear arrangement and isolated position recall the row of huge volcanic peaks that runs parallel with the shore of the Pacific Ocean in Oregon and Washington--Mount Jefferson, Mount Hood, Mount St. Helen's, Mount Tacoma--but these terrestrial volcanoes, except in elevation, are mere pins' heads in the comparison.

In the eastern part of the Mare Fecunditatis lies a pair of relatively small craters named Messier, which possess particular interest because it has been suspected, though not proved, that a change of form has occurred in one or other of the pair. Muller, in the first half of the nineteenth century, represented the two craters as exactly alike in all respects. In 1855 Webb discovered that they are not alike in shape, and that the easternmost one is the larger, and every observer easily sees that Webb's description is correct. Messier is also remarkable for the light streak, often said to resemble a comet's tail, which extends from the larger crater eastward to the shore of the Mare Fecunditatis.

Goclenius and Guttemberg, on the highland between the Mare Fecunditatis and the Mare Nectaris, are intersected and surrounded by clefts, besides being remarkable for their broken and irregular though lofty walls. Guttemberg is forty-five miles and Goclenius twenty-eight miles in diameter. The short mountain range just east of Guttemberg, and bordering a part of the Mare Nectaris on the west, is called the Pyrenees.

The Mare Nectaris, though offering in its appearance no explanation of its toothsome name--perhaps it was regarded as the drinking cup of the Olympian gods--is one of the most singular of the dark lunar plains in its outlines. At the south it ends in a vast semicircular bay, sixty miles across, which is evidently a half-submerged mountain ring. But submerged by what? Not water, but perhaps a sea of lava which has now solidified and forms the floor of the Mare Nectaris. The name of this singular formation is Fracastorius. Elger has an interesting remark about it.

"On the higher portion of the interior, near the center," he says, "is a curious object consisting apparently of four light spots, arranged in a square, with a craterlet in the middle, all of which undergo notable changes of aspect under different phases."

Other writers also call attention to the fine markings, minute craterlets, and apparently changeable spots on the floor of Fracastorius.

We go now to the eastern side of the Mare Nectaris, where we find one of the most stupendous formations in the lunar world, the great mountain ring of Theophilus, noticeably regular in outline and perfect in the completeness of its lofty wall. The circular interior, which contains in the center a group of mountains, one of whose peaks is 6,000 feet high, sinks 10,000 feet below the general level of the moon outside the wall! One of the peaks on the western edge towers more than 18,000 feet above the floor within, while several other peaks attain elevations of 15,000 to 16,000 feet. The diameter of the immense ring, from crest to crest of the wall, is sixty-four miles. Theophilus is especially wonderful on the fifth and sixth days of the moon, when the sun climbs its shining pinnacles and slowly discloses the tremendous chasm that lies within its circles of terrible precipices.

On the southeast Theophilus is connected by extensions of its walls with a shattered ring of vast extent called Cyrillus; and south from Cyrillus, and connected with the same system of broken walls, lies the still larger ring named Catharina, whose half-ruined walls and numerous crater pits present a fascinating spectacle as the shadows retreat before the sunrise advancing across them. These three--Theophilus, Cyrillus, and Catharina--constitute a scene of surpassing magnificence, a glimpse of wonders in another world sufficient to satisfy the most riotous imagination.

South of the Mare Nectaris the huge ring mountain of Piccolomini attracts attention, its massive walls surrounding a floor nearly sixty miles across, and rising in some places to an altitude of nearly 15,000 feet. It should be understood that wherever the height of the mountain wall of such a ring is mentioned, the reference level is that of the interior plain or floor. The elevation, reckoned from the outer side, is always very much less.

The entire region south and east of Theophilus and its great neighbors is marvelously rough and broken. Approaching the center of the moon, we find a system of ringed plains even greater in area than any of those we have yet seen. Hipparchus is nearly a hundred miles long from north to south, and nearly ninety miles broad from east to west. But its walls have been destroyed to such an extent that, after all, it yields in grandeur to a formation like Theophilus.

Albategnius is sixty-five miles across, with peaks from 10,000 to 15,000 feet in height. Sacrobosco is a confused mass of broken and distorted walls. Aliacensis is remarkable for having a peak on the eastern side of its wall which is more than 16,000 feet high. Werner, forty-five miles in diameter, is interesting because under its northeastern wall Muller, some seventy years ago, saw a light spot of astonishing brightness, unmatched in that respect by anything on the moon except the peak of Aristarchus, which we shall see later. This spot seems afterward to have lost brilliance, and the startling suggestion has been made that its original brightness might have been due to its then recent deposit from a little crater that lies in the midst of it. Walter is of gigantic dimensions, about one hundred miles in diameter. Unlike the majority of the ringed plains, it departs widely from a circle. It is yet larger than Walter; but most interesting of all these gigantic formations is Maurolycus, whose diameter exceeds one hundred and fifty miles, and which has walls 13,000 or 14,000 feet high. Yet, astonishing though it may seem, this vast and complicated mass of mountain walls, craters, and peaks, is virtually unseen at full moon, owing to the perpendicularity of the sunlight, which prevents the casting of shadows.

We shall next suppose that another period of about seven days has elapsed, the moon in the meantime reaching its full phase. We refer for guidance to Lunar Chart No. 3. The peculiarity of the northeastern quadrant which immediately strikes the eye is the prevalence of the broad plains called Maria, or "seas." The northern and central parts are occupied by the Mare Imbrium, the "Sea of Showers" or of "Rains," with its dark bay the Sinus tuum, while the eastern half is covered by the vast Oceanus Procellarum, the "Ocean of Storms" or of "Tempests."

Toward the north a conspicuous oval, remarkably dark in hue, immediately attracts our attention. It is the celebrated ringed plain of Plato, about sixty miles in diameter and surrounded by a saw-edged rampart, some of whose pinnacles are 6,000 or 7,000 feet high. Plato is a favorite subject for study by selenographers because of the changes of color which its broad, flat floor undergoes as the sun rises upon it, and also because of the existence of enigmatical spots and streaks whose visibility changes. South of Plato, in the Mare Imbrium, rises a precipitous, isolated peak called Pico, 8,000 feet in height. Its resemblance in situation to the conical mountain Pico in the Azores strikes the observer.

Eastward of Plato a line of highlands, separating the Mare Imbriumfrom the Mare Frigoris, carries the eye to the beautiful semicircular Sinus Iridum, or "Bay of Rainbows." The northwestern extremity of this remarkable bay is guarded by a steep and lofty promontory called Cape Laplace, while the southeastern extremity also has its towering guardian, Cape Heraclides. The latter is interesting for showing, between nine and ten days after full moon, a singularly perfect profile of a woman's face looking out across the Mare

Imbrium. The winding lines, like submerged ridges, delicately marking the floor of the Sinus Iridum and that of the Mare beyond, are beautiful telescopic objects. The "bay" is about one hundred and thirty-five miles long by eighty-four broad.

The Mare Imbrium, covering 340,000 square miles, is sparingly dotted over with craters. All of the more conspicuous of them are indicated in the chart. The smaller ones, like Caroline Herschel, Helicon, Leverrier, Delisle, etc., vary from eight to twelve miles in diameter. Lambert is seventeen miles in diameter, and Euler nineteen, while Timocharis is twenty-three miles broad and 7,000 feet deep below its walls, which rise only 3,000 feet above the surface of the Mare.

Toward the eastern border of the sea, south of the Harbinger Mountains, we find a most remarkable object, the mountain ring, or crater plain, called Aristarchus. This ring is not quite thirty miles in diameter, but there is nothing on the moon that can compare with it in dazzling brilliance. The central peak, 1,200 or 1,300 feet high, gleams like a mountain of crusted snow, or as if it were composed of a mass of fresh-broken white metal, or of compacted crystals. Part of the inner slope of the east wall is equally brilliant. In fact, so much light is poured out of the circumvallation that the eye is partially blinded, and unable distinctly to see the details of the interior. No satisfactory explanation of the extraordinary reflecting power of Aristarchus has ever been offered. Its neighbor toward the east, Herodotus, is somewhat smaller and not remarkably bright, but it derives great interest from the fact that out of a breach in its northern wall issues a vast cleft, or chasm, which winds away for nearly a hundred miles across the floor of the Mare, making an abrupt turn when it reaches the foot of the Harbinger Mountains.

The comparatively small crater, Lichtenberg, near the northeastern limb of the moon, is interesting because Muller used to see in its neighborhood a pale-red tint which has not been noticed since his day.

Returning to the western side of the quadrant represented in Lunar Chart No. 3, we see the broad and beautifully regular ringed plain of Archimedes, fifty miles in diameter and 4,000 feet deep.

A number of clefts extend between the mountainous neighborhood of

Archimedes and the feet of the gigantic Apennine Mountains on the southwest. The little double crater, Beer, between Archimedes and Timocharis, is very bright.

The Apennines extend about four hundred and eighty miles in a northwesterly and southeasterly direction. One of their peaks near the southern end of the range, Mount Huygens, is at least 18,000 feet high, and the black silhouettes of their sharp-pointed shadows thrown upon the smooth floor of the Mare Imbrium about the time of first quarter present a spectacle as beautiful as it is unique. The Apennines end at the southeast in the ring mountain, Eratosthenes, thirty-eight miles across and very deep, one of its encircling chain of peaks rising 16,000 feet above the floor, and about half that height above the level of the Mare Imbrium. The shadows cast by Eratosthenes at sunrise are magnificent.

And now we come to one of the supreme spectacles of the moon, the immense ring or crater mountain Copernicus. This is generally regarded as the grandest object that the telescope reveals on the earth's satellite. It is about fifty-six miles across, and its interior falls to a depth of 8,000 feet below the Mare Imbrium. Its broad wall, composed of circle within circle of ridges, terraces, and precipices, rises on the east about 12,000 feet above the floor. On the inner side the slopes are very steep, cliff falling below cliff, until the bottom of the fearful abyss is attained. To descend those precipices and reach the depressed floor of Copernicus would be a memorable feat for a mountaineer. In the center of the floor rises a complicated mountain mass about 2,400 feet high. All around Copernicus the surface of the moon is dotted with countless little crater pits, and splashed with whitish streaks. Northward lie the Carpathian Mountains, terminating on the east in Tobias Mayer, a ring mountain more than twenty miles across. The mountain ring Kepler, which is also the center of a great system of whitish streaks and splashes, is twenty-two miles in diameter, and notably brilliant.

Finally, we turn to the southeastern quadrant of the moon, represented in Lunar Chart No. 4. The broad, dark expanse extending from the north is the Mare Nubium on the west and the Oceanus Procellarum on the east. Toward the southeast appears the notably dark, rounded area of the Mare Humorum inclosed by highlands and rings. We begin with the range of vast inclosures running southward near the central meridian, and starting with Ptolemy's, a walled plain one hundred and fifteen miles in its greatest diameter and covering an area considerably exceeding that of the State of Massachusetts. Its neighbor toward the south, Alphonsus, is eighty-three miles across. Next comes Arzachel, more than sixty-five miles in diameter. Thebit, more than thirty miles across, is very deep. East of Thebit lies the celebrated "lunar railroad," a straight, isolated wall about five hundred feet high and sixty-five miles long, dividing at its southern end into a number of curious branches, forming the buttresses of a low mountain. Purbach is sixty miles broad, and south of that comes a wonderful region where the ring mountains Hell, Ball, Lexell, and others, more or less connected with walls, inclose an area even larger than Ptolemy's, but which, not being so distinctly bordered as some of the other inclosed plains, bears no distinctive name.

The next conspicuous object toward the south ranks with Copernicus among the grandest of all lunar phenomena--the ring, or crater, Tycho. It is about fifty-four miles in diameter and some points on its wall rise 17,000 feet above the interior. In the center is a bright mountain peak 5,000 feet high. But wonderful as are the details of its mountain ring, the chief attraction of Tycho is its manifest relation to the mysterious bright rays heretofore referred to, which extend far across the surface of the moon in all directions, and of which it is the center. The streaks about Copernicus are short and confused, constituting rather a splash than a regular system of rays; but those emanating from Tycho are very long, regular, comparatively narrow, and form arcs of great circles which stretch away for hundreds of miles, allowing no obstacle to interrupt their course.

Southwest of Tycho lies the vast ringed plain of Maginus, a hundred miles broad and very wonderful to look upon, with its labyrinth of formations, when the sun slopes across it, and yet, like Maurolycus, invisible under a vertical illumination. "The full moon," to use Muller's picturesque expression, "knows no Maginus." Still larger and yet more splendid is Clavius, which exceeds one hundred and forty miles in diameter and covers 16,000 square miles of ground within its fringing walls, which carry some of the loftiest peaks on the moon, several attaining 17,000 feet. The floor is deeply depressed, so that an inhabitant of this singular inclosure, larger than Massachusetts, Connecticut, and Rhode Island combined, would dwell in land sunk two miles below the general level of the world about him.

In the neighborhood of the south pole lies the large walled plain of Newton, whose interior is the deepest known depression on the moon. It is so deep that the sunshine never touches the larger part of the floor of the inner abyss, and a peak on its eastern wall rises 24,000 feet sheer above the tremendous pit. Other enormous walled plains are Longomontanus, Wilhelm I, Schiller, Bailly, and Schickard. The latter is one hundred and thirty-four miles long and bordered by a ring varying from 4,000 to 9,000 feet in height. Wargentin, the oval close to the moon's southeast limb, beyond Schickard, is a unique formation in that, instead of its interior being sunk below the general level, it is elevated above it. It has been suggested that this peculiarity is due to the fact that the floor of Wargentin was formed by inflation from below, and that it has cooled and solidified in the shape of a gigantic dome arched over an immense cavity beneath. A dome of such dimensions, however, could not retain its form unless partly supported from beneath.

Hainzel is interesting from its curious outline; Cichus for the huge yawning crater on its eastern wall; Capuanus for a brilliant shining crater also on its eastern wall; and Mercator for possessing bright craters on both its east and its west walls. Vitello has a bright central mountain and gains conspicuousness from its position at the edge of the dark Mare Humorum. Agatharchides is the broken remnant of a great ring mountain. Gassendi, an extremely beautiful object, is about fifty-five miles across. It is encircled with broken walls, craters and bright points, and altogether presents a very splendid appearance about the eleventh day of the moon's age.

Letronne is a half-submerged ring, at the southern end of the Oceanus Procellarum, which recalls Fracastorius in the western lunar hemisphere. It lies, however, ten degrees nearer the equator than Fracastorius. Billy is a mountain ring whose interior seems to have been submerged by the dark substance of the Oceanus Procellarum, although its walls have remained intact. Mersenius is a very conspicuous ring, forty miles in diameter, east of the Mare Humorum. Vieta, fifty miles across, is also a fine object. Grimaldi, a huge dusky oval, is nearly one hundred and fifty miles in its greatest length. The ring mountain Landsberg, on the equator, and near the center of the visible eastern hemisphere, is worth watching because Elger noticed changes of color in its interior in 1888.

Bullialdus, in the midst of the Mare Nubium, is a very conspicuous and beautiful ring mountain about thirty-eight miles in diameter, with walls 8,000 feet high above the interior.

Those who wish to see the lunar mountains in all their varying aspects will not content themselves with views obtained during the advance of the sunlight from west to east, between "new moon" and "full moon," but will continue their observations during the retreat of the sunlight from east to west, after the full phase is passed.

It is evident that the hemisphere of the moon which is forever turned away from the earth is quite as marvelous in its features as the part that we see. It will be noticed that the entire circle of the moon's limb, with insignificant interruptions, is mountainous. Possibly the invisible side of our satellite contains yet grander peaks and crater mountains than any that our telescopes can reach. This probability is increased by the fact that the loftiest known mountain on the moon is never seen except in silhouette. It is a member of a great chain that breaks the lunar limb west of the south pole, and that is called the Leibnitz Mountains. The particular peak referred to is said by some authorities to exceed 30,000 feet in height. Other great ranges seen only in profile are thel Mountains on the limb behind the ring plain Bailly, the Cordilleras, east of Eichstadt, and the D'Alembert Mountains beyond Grimaldi. The profile of these great mountains is particularly fine when they are seen during an eclipse of the sun. Then, with the disk of the sun for a background, they stand out with startling distinctness.

THE SUN

When the sun is covered with spots it becomes a most interesting object for telescopic study. Every amateur's telescope should be provided with apparatus for viewing the sun. A dark shade glass is not sufficient and not safe. What is known as a solar prism, consisting of two solid prisms of glass, cemented together in a brass box which carries a short tube for the eyepiece, and reflecting an image of the sun from their plane of junction--while the major remnant of light and heat passes directly through them and escapes from an opening provided for the purpose--serves very well. Better and more costly is an apparatus called a helioscope, constructed on the principle of polarization and provided with prisms and reflectors which enable the observer, by proper adjustment, to govern very exactly and delicately the amount of light that passes into the eyepiece.

Furnished with an apparatus of this description we can employ either a three-, four-, or five-inch glass upon the sun with much satisfaction. For the amateur's purposes the sun is only specially interesting when it is spotted. The first years of the twentieth century will behold a gradual growth in the number and size of the solar spots as those years happen to coincide with the increasing phase of the sun-spot period. Large sun spots and groups of spots often present an immense amount of detail which tasks the skill of the draughtsman to represent it. But a little practice will enable one to produce very good representations of sun spots, as well as of the whitish patches called facul?by which they are frequently surrounded.

For the simple purpose of exhibiting the spotted face of the sun without much magnifying power, a telescope may be used to project the solar image on a white sheet or screen. If the experiment is tried in a room, a little ingenuity will enable the observer to arrange a curtain covering the window used, in such a way as to exclude all the light except that which comes through the telescope. Then, by placing a sheet of paper or a drawing board before the eyepiece and focusing the image of the sun upon it, very good results may be obtained.

If one has a permanent mounting and a driving clock, a small spectroscope may be attached, for solar observations, even to a telescope of only four or five inches aperture, and with its aid most interesting views may be obtained of the wonderful red hydrogen flames that frequently appear at the edge of the solar disk.

CHAPTER X

ARE THERE PLANETS AMONG THE STARS?

"... And if there should be Worlds greater than thine own, inhabited By greater things, and they themselves far more In number than the dust of thy dull earth, What wouldst thou think?"--BYRON'S CAIN.

This always interesting question has lately been revived in a startling manner by discoveries that have seemed to reach almost deep enough to touch its solution. The following sentences, from the pen of Dr. T. J. J. See, of the Lowell Observatory, are very significant from this point of view:

"Our observations during 1896-'97 have certainly disclosed stars more difficult than any which astronomers had seen before. Among these obscure objects about half a dozen are truly wonderful, in that they seem to be dark, almost black in color, and apparently are shining by a dull reflected light. It is unlikely that they will prove to be self-luminous. If they should turn out dark bodies in fact, shining only by the reflected light of the stars around which they revolve, we should have the first case of planets--dark bodies--noticed among the fixed stars."

Of course, Dr. See has no reference in this statement to the immense dark bodies which, in recent years, have been discovered by spectroscopic methods revolving around some of the visible stars, although invisible themselves. The obscure objects that he describes belong to a different class, and might be likened, except perhaps in magnitude, to the companion of Sirius, which, though a light-giving body, exhibits nevertheless a singular defect of luminosity in relation to its mass. Sirius has only twice the mass, but ten thousand times the luminosity, of its strange companion! Yet the latter is evidently rather a faint, or partially extinguished, sun than an opaque body shining only with light borrowed from its dazzling neighbor. The objects seen by Dr. See, on the contrary, are "apparently shining by a dull reflected light."

If, however (as he evidently thinks is probable), these objects should prove to be really non-luminous, it would not follow that they are to be regarded as more like the planets of the solar system than like the dark companions of certain other stars. A planet, in the sense which we attach to the word, can not be comparable in mass and size with the sun around which it revolves. The sun is a thousand times larger than the greatest of its attendant planets, Jupiter, and more than a million times larger than the earth. It is extremely doubtful whether the relation of sun and planet could exist between two bodies of anything like equal size, or even if one exceeded the other many times in magnitude. It is only when the difference is so great that the smaller of the two bodies is insignificant in comparison with the larger, that the former could become a cool, life-bearing globe, nourished by the beneficent rays of its organic comrade and master.

Judged by our terrestrial experience, which is all we have to go by, the magnitude of a planet, if it is to bear life resembling that of the earth, is limited by other considerations. Even Jupiter, which, as far as our knowledge extends, represents the extreme limit of great planetary size, may be too large ever to become the abode of living beings of a high organization. The force of gravitation on the surface of Jupiter exceeds that on the earth's surface as 2.64 to 1. Considering the effects of this on the weight and motion of bodies, the density of the atmosphere, etc., it is evident that Jupiter would, to say the very least, be an exceedingly uncomfortable place of abode for beings resembling ourselves. But Jupiter, if it is ever to become a solid, rocky globe like ours, must shrink enormously in volume, since its density is only 0.24 as compared with the earth. Now, the surface gravity of a planet depends on its mass and its radius, being directly as the former and inversely as the square of the latter. But in shrinking Jupiter will lose none of its mass, although its radius will become much smaller. The force of gravity will consequently increase on its surface as the planet gets smaller and more dense.

The present mean diameter of Jupiter is 86,500 miles, while its mass exceeds that of the earth in the ratio of 316 to 1. Suppose Jupiter shrunk to three quarters of its present diameter, or 64,800 miles, then its surface gravity would exceed the earth's nearly five times. With one half its present diameter the surface gravity would become more than ten times that of the earth. On such a planet a man's bones would snap beneath his weight, even granting that he could remain upright at all! It would seem, then, that, unless we are to abandon terrestrial analogies altogether and "go it blind," we must set an upper limit to the magnitude of a habitable planet, and that Jupiter represents such upper limit, if, indeed, he does not transcend it.

The question then becomes, Can the faint objects seen by Dr. See and his fellow-observers, in the near neighborhood of certain stars, be planets in the sense just described, or are they necessarily far greater in magnitude than the largest planet, in the accepted sense of that word, which can be admitted into the category--viz., the planet Jupiter? This resolves itself into another question: At what distance would Jupiter be visible with a powerful telescope, supposing it to receive from a neighboring star an amount of illumination not less than that which it gets from the sun? To be sure, we do not know how far away the faint objects described by Dr. See are; but, at any rate, we can safely assume that they are at the distance of the nearest stars, say somewhere about three hundred thousand times the earth's distance from the sun. The sun itself removed to that distance would appear to our eyes only as a star of the first magnitude. But Zolner has shown that the sun exceeds Jupiter in brilliancy 5,472,000,000 times. Seen from equal distances, however, the ratio would be about 218,000,000 to 1. This would be the ratio of their light if both sun and Jupiter could be removed to about the distance of the nearest stars. Since the sun would then be only as bright as one of the stars of the first magnitude, and since Jupiter would be 218,000,000 times less brilliant, it is evident that the latter would not be visible at all. The faintest stars that the most powerful telescopes are able to show probably do not fall below the sixteenth or, at the most, the seventeenth magnitude. But a seventeenth-magnitude star is only between two and three million times fainter than the sun would appear at the distance above supposed, while, as we have seen, Jupiter would be more than two hundred million times fainter than the sun.

To put it in another way: Jupiter, at the distance of the nearest stars, would be not far from one hundred times less bright than the faintest star which the largest telescope is just able, under the most exquisite conditions, to glimpse. To see a star so faint as that would require an object-glass of a diameter half as great as the length of the tube of the Lick telescope, or say thirty feet!

Of course, Jupiter might be more brilliantly illuminated by a brighter star than the sun; but, granting that, it still would not be visible at such a distance, even if we neglect the well-known concealing or blinding effect of the rays of a bright star when the observer is trying to view a faint one close to it. Clearly, then, the obscure objects seen by Dr. See near some of the stars, if they really are bodies visible only by light reflected from their surfaces, must be enormously larger than the planet Jupiter, and can not, accordingly, be admitted into the category of planets proper, whatever else they may be.

Perhaps they are extreme cases of what we see in the system of Sirius--i.e., a brilliant star with a companion which has ceased to shine as a star while retaining its bulk. Such bodies may be called planets in that they only shine by reflected light, and that they are attached to a brilliant sun; but the part that they play in their systems is not strictly planetary. Owing to their great mass they bear such sway over their shining companions as none of our planets, nor all of them combined, can exercise; and for the same reason they can not, except in a dream, be imagined to possess that which, in our eyes, must always be the capital feature of a planet, rendering it in the highest degree interesting wherever it may be found--sentient life.

It does not follow, however, that there are no real planetary bodies revolving around the stars. As Dr. See himself remarks, such insignificant bodies as our planets could not be seen at the distance of the fixed stars, "even if the power of our telescopes were increased a hundredfold, and consequently no such systems are known."

This brings me to another branch of the subject. In the same article from which I have already quoted (Recent Discoveries respecting the Origin of the Universe, Atlantic Monthly, vol. lxxx, pages 484-492), Dr. See sets forth the main results of his well-known studies on the origin of the double and multiple star systems. He finds that the stellar systems differ from the solar system markedly in two respects, which he thus describes:

"1. The orbits are highly eccentric; on the average twelve times more elongated than those of the planets and satellites.

"2. The components of the stellar systems are frequently equal and always comparable in mass, whereas our satellites are insignificant compared to their planets, and the planets are equally small compared to the sun."

These peculiarities of the star systems Dr. See ascribes to the effect of "tidal friction," the double stars having had their birth through fission of original fluid masses (just as the moon, according to George Darwin's theory, was born from the earth), and the reaction of tidal friction having not only driven them gradually farther apart but rendered their orbits more and more eccentric. This manner of evolution of a stellar system Dr. See contrasts with Laplace's hypothesis of the origin of the planetary system through the successive separation of rings from the periphery of the contracting solar nebula, and the gradual breaking up of those rings and their aggregation into spherical masses or planets. While not denying that the process imagined by Laplace may have taken place in our system, he discovers no evidence of its occurrence among the double stars, and this leads him to the following statement, in which believers in the old theological doctrine that the earth is the sole center of mortal life and of divine care would have found much comfort:

"It is very singular that no visible system yet discerned has any resemblance to the orderly and beautiful system in which we live; and one is thus led to think that probably our system is unique in its character. At least it is unique among all known systems."

If we grant that the solar system is the only one in which small planets exist revolving around their sun in nearly circular orbits, then indeed we seem to have closed all the outer universe against such beings as the inhabitants of the earth. Beyond the sun's domain only whirling stars, coupled in eccentric orbits, dark stars, some of them, but no planets--in short a wilderness, full of all energies except those of sentient life! This is not a pleasing picture, and I do not think we are driven to contemplate it. Beyond doubt, Dr. See is right in concluding that double and multiple star systems, with their components all of magnitudes comparable among themselves, revolving in exceedingly eccentric orbits under the stress of mutual gravitation, bear no resemblance to the orderly system of our sun with its attendant worlds. And it is not easy to imagine that the respective members of such systems could themselves be the centers of minor systems of planets, on account of the perturbing influences to which the orbits of such minor systems would be subjected.

But the double and multiple stars, numerous though they be, are outnumbered a hundred to one by the single stars which shine alone as our sun does. What reason can we have, then, for excluding these single stars, constituting as they do the vast majority of the celestial host, from a similarity to the sun in respect to the manner of their evolution from the original nebulous condition? These stars exhibit no companions, such planetary attendants as they may have lying, on account of their minuteness, far beyond the reach of our most powerful instruments. But since they vastly outnumber the binary and multiple systems, and since they resemble the sun in having no large attendants, should we be justified, after all, in regarding our system as "unique"? It is true we do not know, by visual evidence, that the single stars have planets, but we find planets attending the only representative of that class of stars that we are able to approach closely--the sun--and we know that the existence of those planets is no mere accident, but the result of the operation of physical laws which must hold good in every instance of nebular condensation.

Two different methods are presented in which a rotating and contracting nebula may shape itself into a stellar or planetary system. The first is that described by Laplace, and generally accepted as the probable manner of origin of the solar system--viz., the separation of rings from the condensing mass, and the subsequent transformation of the rings into planets. The planet Saturn is frequently referred to as an instance of the operation of this law, in which the evolution has been arrested after the separation of the rings, the latter having retained the ring form instead of breaking and collecting into globes, forming in this case rings of meteorites, and reminding us of the comparatively scattered rings of asteroids surrounding the sun between the orbits of Mars and Jupiter. This Laplacean process Dr. See regards as theoretically possible, but apparently he thinks that if it took place it was confined to our system.

The other method is that of the separation of the original rotating mass into two nearly equal parts. The mechanical possibility of such a process has been proved, mathematically, by Poincar?and Darwin. This, Dr. See thinks, is the method which has prevailed among the stars, and prevailed to such a degree as to make the solar system, formed by the ring method, probably a unique phenomenon in the universe.

Is it not more probable that both methods have been in operation, and that, in fact, the ring method has operated more frequently than the other? If not, why do the single stars so enormously outnumber the double ones? It is of the essence of the fission process that the resulting masses should be comparable in size. If, then, that process has prevailed in the stellar universe to the practical exclusion of the other, there should be very few single stars; whereas, as a matter of fact, the immense majority of the stars are single. And, remembering that the sun viewed from stellar distances would appear unattended by subsidiary bodies, are we not justified in concluding that its origin is a type of the origin of the other single stars?

While it is, as I have remarked, of the essence of the fission process that the resulting parts of the divided mass should be comparable in magnitude, it is equally of the essence of the ring, or Laplacean process, that the bodies separated from the original mass should be comparatively insignificant in magnitude.

As to the coexistence of the two processes, we have, perhaps, an example in the solar system itself. Darwin's demonstration of the possible birth of the moon from the earth, through fission and tidal friction, does not apply to the satellites attending the other planets. The moon is relatively a large body, comparable in that respect with the earth, while the satellites of Jupiter and Saturn, for instance, are relatively small. But in the case of Saturn there is visible evidence that the ring process of satellite formation has prevailed. The existing rings have not broken up, but their very existence is a testimony of the origin of the satellites exterior to them from other rings which did break up. Thus we need not go as far away as the stars in order to find instances illustrating both the methods of nebular evolution that we have been dealing with.

The conclusion, then, seems to be that we are not justified in assuming that the solar system is unique simply because it differs widely from the double and multiple star systems; and that we should rather regard it as probable that the vast multitude of stars which do not appear, when viewed with the telescope, or studied by spectroscopic methods, to have any attendants comparable with themselves in magnitude, have originated in a manner resembling that of the sun's origin, and may be the centers of true planetary systems like ours. The argument, I think, goes further than to show the mere possibility of the existence of such planetary systems surrounding the single stars. If those stars did not originate in a manner quite unlike the origin of the sun, then the existence of planets in their neighborhood is almost a foregone conclusion, for the sun could hardly have passed through the process of formation out of a rotating nebula without evolving planets during its contraction. And so, notwithstanding the eccentricities of the double stars, we may still cherish the belief that there are eyes to see and minds to think out in celestial space.

###

www.ingramcontent.com/pod-product-compliance
Lightning Source LLC
Chambersburg PA
CBHW070904180526
45168CB00005B/1927

* 9 7 8 1 5 1 4 8 0 0 1 4 0 *